Liquid Crystals

LIQUID CRYSTALS

Nature's Delicate Phase of Matter

BY PETER J. COLLINGS

Adam Hilger, Bristol

British Library Cataloguing in Publication Data

Collings, Peter J.
Liquid crystals.
1. Liquid crystals
I. Title
530.429

ISBN 0-7503-0054-X
ISBN 0-7503-0055-8 pbk

Published under the Adam Hilger imprint by IOP Publishing Ltd
Techno House, Redcliffe Way, Bristol BS1 6NX, England

Printed in the United States of America by Princeton University Press,
Princeton, New Jersey

To my parents, for their caring guidance
and support in spite of my insensitivity;
to my wife, whose love and encouragement
taught me to be otherwise;
and to my children, who appropriately
are starting the cycle all over again

Contents

List of Plates

Following p. 114

Preface

I have heard it argued, usually by nonscientists, that to study nature using the methods of science tends to diminish the glory of the natural world in which we live. The claim is that nature is somehow "demystified" and simplified by science. I doubt very much if a scientist would ever argue in this way. Any person who has seriously examined what we know about the inner workings of a cell or carefully considered the theory of matter during the first minutes of our universe can only stand in awe of nature's elaborate yet functional design. Even old discoveries never seem to lose their charm. Each new hypothesis or experimental result has a way of affecting everything that came before it, sometimes in a dramatic way and often only in a subtle way. Our view of nature is constantly in flux, with new revelations occurring in both expected and unexpected areas. Scientists rest only momentarily to celebrate knowing something new about nature, because this new "worldview" always changes what they thought they knew and forces them to ask questions they never would have thought to ask. This scientific perspective on the natural world is both beautiful and glorious, but not in a mysterious way. It is more analogous to the relationship between two good friends. Rather than being a detriment to the relationship, the increasing familiarity in a friendship allows both people to develop in ways not possible individually. Such is the case with science and nature.

Nothing exemplifies these ideas better for me than my studies involving liquid crystals. I have been engaged in liquid crystal research for over sixteen years, and as a result know much more about them than when I started. And yet, they never cease to amaze me. I recall very vividly the first time I actually saw a substance form a liquid crystal phase. I had been learning about liquid crystals for a few months, and understood many of their properties. Yet observing what I had only previously imagined added meaning to what I had learned and made me want to learn more. The truth is that I continue to have those same feelings at times in the laboratory. It still happens once in a while as I watch a substance form a liquid crystal phase under the microscope, but more often I get that feeling when an experiment reveals that the liquid crystal does not behave as our theories predict. All I can think about is why, and I want to learn more.

In writing this book, I could not avoid attempting to convey some of these feelings of wonder and beauty. I have also tried to include as much background information as necessary, so a reader need not consult other references in order to understand the concepts. Hopefully this will allow the reader to learn more about liquid crystals, while at the same time finding enjoyment in understanding this beautiful phase of matter. I have also tried to be as comprehensive as possible, covering all the important areas in biology, chemistry, and physics. In doing this, I have tried to keep the scientific level consistent throughout the book. A reader familiar with the basic ideas of biology, chemistry, and physics should have little difficulty in following the explanations and arguments. This book includes a short chapter on the discovery and early history of liquid crystals, as well as a chapter that discusses some of the more modern and theoretical research being performed at this time. Two chapters discuss many of the important technical applications of liquid crystals, with appropriate emphasis on the liquid crystal display. Finally, ample space is devoted to the exciting work being done with polymer liquid crystals, liquid crystal emulsions, and liquid crystals in biological systems.

Acknowledgments

Many people have provided important assistance during the writing of this book. I thank J. William Doane and Hassan Hakemi for encouraging me to write a book of this type; Diane Collings, Franklin Miller, Jr., Joan Slonczewski, and Owen York for critically reading parts of the manuscript; Thomas Greenslade, Jr., for reproducing some of the pictures; and Brandon Collings for drawing all the figures. I also acknowledge that the initial work on this book was supported by a Summer Research Fellowship from Kenyon College. Finally, I thank the following people for providing pictures for the book: Dr. W. Becker (E. Merck Company); Dr. P. Crooker; Dr. J. Goodby; Dr. H. Kitzerow; Mr. E. Juge (Tandy Corporation/Radio Shack); Dr. M. Neubert; Dr. C. Nocka (F. Hoffmann-LaRoche Company); Ms. M. Noordhoff (Taliq Corporation); Dr. E. Samulski; Mr. L. Santacaterina (Qmax Technology Group); Dr. H. Stegemeyer; and Dr. N. Vaz (General Motors Corporation).

Liquid Crystals

CHAPTER 1

What Are Liquid Crystals?

A famous scientist once remarked that he found liquid crystals to be quite mysterious. Certainly the name itself is confusing. After all, how can something be both liquid and crystalline? We shall see in this chapter that in some cases a very interesting compromise spontaneously occurs in the natural world, and that the name used to describe the result of this compromise is quite appropriate. It will also be evident that liquid crystals come in a number of forms, creating an area of study rich in new and exciting phenomena.

States of Matter

We are all aware of the fact that many substances can exist in more than one state of matter. The most familiar example is water, which is a solid below 0°C (32°F), a liquid between 0° and 100°C (212°F), and a gas above 100°C. Solids, liquids, and gases are the most common states of matter, but in spite of what many children learn in school, they are not the only states of matter.

These three common states of matter are different from each other because the molecules in each state possess different amounts of order. The solid state consists of a more or less rigid arrangement of molecules because each molecule occupies a certain place in the arrangement and remains there. Not only are the molecules constrained to occupy a specific position, but they are also oriented in a specific way. The molecules might vibrate a bit, but on average they constantly maintain this highly ordered arrangement. There are large attractive forces holding the molecules of a solid in place, because the arrangement causes the forces between individual molecules to add together. It therefore takes large external forces to disrupt the structure, so solids are hard and difficult to deform.

The liquid state is quite different in that the molecules neither occupy a specific average position nor remain oriented in a particular way. The molecules are free to diffuse about in a random fashion, constantly bumping into one another and abruptly changing their direction of motion. The amount of order in a liquid is therefore much less than in a solid. Attractive forces still exist in a liquid, but the random motion of the molecules does not allow the forces between individual molecules to add together. The result is that the forces holding liquid molecules together tend to be much weaker than the forces in solids. These forces are strong enough, however, to keep the molecules fairly close to each other. A liquid, therefore, maintains a constant density, even though it takes the shape of its container. The lack of a rigid arrangement allows liquids to be deformed quite easily. Liquids flow and will change their shape in response to weak outside forces. The fact that the molecules are held close together is evident when one tries to compress a liquid. Liquids are difficult to compress, a characteristic that is put to successful use in the hydraulic systems of automobile brakes and earth-moving equipment.

In the gas state, the more chaotic motion of the molecules causes the attractive forces between them to add together even less than in the liquid state. The amount of order is therefore less than in liquids and the forces are not strong enough to hold the molecules close together. The molecules move about in the same fashion as in liquids, but they eventually spread evenly throughout the container no matter how large it is. Thus, the liquid and gas states are very similar; the motion of the molecules is chaotic and random in both states. But in the liquid state the molecules maintain a specific average distance between one another, while in the gas state the average intermolecular distance is determined by the number of molecules and the size of the container. A gas can be deformed even more easily than a liquid. In fact, a gas can be significantly compressed since it takes much less force to move the molecules a little closer together.

I should mention that the description I have given for the solid state is really appropriate for a crystalline solid. Some substances in the liquid state change into an amorphous solid rather than a crystalline solid when cooled. The molecules in an amorphous solid are fixed in place, but there is no overall pattern to their arrangement. The molecules are arranged more or less randomly (similar to a snapshot taken in a liquid) but unlike in the liquid state, they do not diffuse throughout the substance. Some of the substances discussed in this book possess crystalline solid states while others possess amorphous solid states.

To understand why different substances form certain phases, we must consider the effects due to temperature. Temperature is a measure of the random motion of molecules. The higher the temperature, the more the molecules are moving and vibrating in a random way. Since the attractive forces between the molecules of a substance in a certain state of matter do not change with temperature (although the random motion of molecules does increase with temperature), the ability of the attractive intermolecular forces to keep the molecules ordered in any way must decrease as the temperature increases. Consider water for example. Below 0°C the attractive forces between the water molecules when arranged so rigidly are strong enough to hold the molecules firmly in place, even though they all possess random motion due to the temperature. Above 0°C, however, the random motion becomes too violent and the intermolecular forces cannot hold the molecules in place, causing the solid to melt. The random motion causes the molecules to wander around, which reduces the forces holding them together because the forces between molecules no longer add together as in the solid state. These forces, however, are still great enough to keep the molecules from wandering completely away from each other. The liquid takes up a specific amount of space. At a temperature above 100°C the random motion is so violent that the attractive forces are not even capable of keeping the molecules next to one another. The water is now in the gas phase, and the intermolecular forces are even weaker than in the liquid phase, because the forces between molecules add together even less. The molecules spread out to fill the available volume.

I can now make a generalization from this discussion concerning water. Every substance possesses intermolecular forces of some kind, so at any temperature, a substance exists in a specific state of matter. Scientists also use the word *phase* to describe a specific state of matter, and refer to the specific state as the stable phase at that temperature. For example, water at 20°C (68°F) is stable in the liquid phase. Carbon dioxide is stable in the gas phase at this temperature, and salt (sodium chloride) is stable in the solid phase. The intermolecular forces are different in these three substances, in each case causing a different phase to be stable at the same temperature.

When the temperature is changed so that the phase is no longer stable, the substance changes phase. The change of phase occurs at a precise temperature because at that temperature the ability of the intermolecular forces to cause that phase to exist is no longer sufficient. We say a *phase transition* occurs at that temperature. The important change that takes place at a phase transition is the amount of order among the molecules of the substance.

In most cases the process is reversible. A liquid to solid phase transition occurs at 0°C for water if it is being cooled. The fact that any substance has phases that are stable over certain temperature ranges allows us to draw a single diagram for each substance summarizing its phase behavior. A horizontal line represents temperature, with higher temperature to the right. Phase transitions are indicated by short vertical lines at the proper temperature, and the names of the stable phases are written between the appropriate phase transition lines. The diagram for water is shown in figure 1.1. A schematic diagram of a molecule of water is also depicted in the figure. In diagrams of molecules, letters are used to represent the atoms with short straight lines representing the fact that two atoms share electrons and therefore are bound together. This convention is used throughout the book.

Before I proceed further, let me digress and mention a fourth state of matter called the *plasma* phase. It has nothing to do with liquid crystals, but it is a true state of matter just as the solid, liquid, and gas states are. If a substance is heated to a high enough temperature

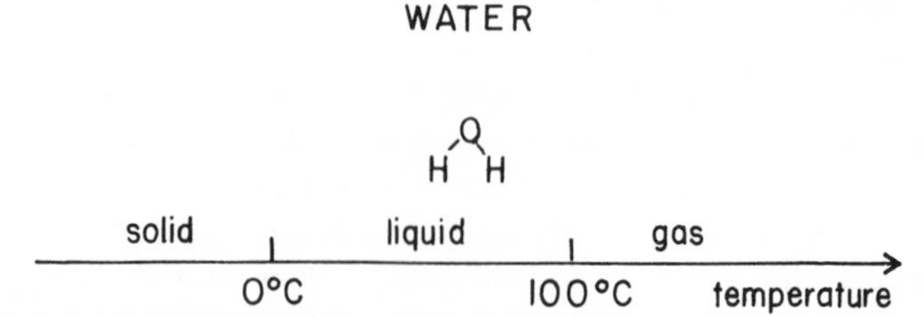

Fig. 1.1 Phase diagram of water. The stable phases of water are given above the temperature axis, with the phase transition temperatures shown below the axis.

(and I do mean very high), the random motion becomes so violent that electrons which are normally bound to the atoms get knocked off and cannot recombine. This phase of matter is composed of positively charged ions (atoms that have lost one or more electrons) and negatively charged electrons, which normally attract each other so strongly that the ions and electrons bind together. The temperature is so high, however, that the substance exists in this state with unbound ions and electrons. It is a new phase of matter, one that normally exists in and around stars. Scientists presently create a plasma in their experiments on nuclear fusion. The temperatures of these plasmas are so high that scientists continue to learn new things about how matter behaves under these conditions.

The Liquid Crystal Phase

I will now complicate this picture one more time and discuss liquid crystals. Instead of doing an experiment with water, let us imagine that we have a substance called cholesteryl myristate and wish to investigate its phases. Cholesteryl myristate is a complicated molecule, composed mostly of carbon and hydrogen atoms. It is a fairly common substance, however, and can be found in our cell membranes and also in those nasty deposits that cause hardening of our arteries. At room temperature (20°C) it is in the solid phase, so we heat it up slowly and observe what takes place. At 71°C the solid melts, but the resulting ''liquid'' is very cloudy and not like other liquids such as water, alcohol, or cooking oil. If we keep raising the temperature, we notice that another change takes place at 85°C; here the cloudy ''liquid'' turns clear, now looking like other familiar liquids. We continue to heat up the substance; nothing further happens, and we reach the maximum temperature of our apparatus (200°C).

Based on what I have discussed, you may be thinking that these abrupt changes at a specific temperature must be phase transitions. If so, the cloudy ''liquid'' must represent a phase different from both the solid and liquid phases. This idea is indeed correct, and the phase is called the *liquid crystal* phase. It is a fluid phase in that a liquid crystal flows and will take the shape of its container. Its cloudiness, however, indicates that it differs from liquids in some fundamental way. It was no simple job for scientists of the late nineteenth and early twentieth centuries to put together a proper conception of this new state of matter. But through lots of experimentation and some creative thinking, a useful picture of the liquid crystal phase began to emerge. Before going any further, let us draw a diagram showing the phases of cholesteryl myristate (see figure 1.2). I am certain a gas phase of cholesteryl myristate exists at some high temperature, but the molecules begin to break apart under these extreme conditions. Therefore, it might be impossible to obtain a gas phase of the pure compound.

The molecules in a solid are constrained to occupy only certain positions. We describe this condition by saying that the solid phase possesses *positional order*. In addition, the molecules in these specific positions are also constrained in the ways they orient themselves with respect to one another. We say the solid phase also possesses *orientational order*. When a solid melts to a liquid, both types of order are lost completely; the molecules move and tumble randomly. When a solid melts to a liquid crystal, however, the positional order may be lost although some of the orientational order remains. The molecules in the liquid crystal phase are free to move about in much the same fashion as in a liquid; but as they do so they tend to remain oriented in a certain direction. This orientational order is not nearly as perfect as in a solid; in fact, the molecules of a liquid crystal spend only a little more time pointing along the direction of orientation than along some other direction. Still this partial alignment does represent a degree of order not present in liquids and thus requires that we call this condition a new phase or state of matter. Figure 1.3 illustrates the order present in the solid, liquid crystal, and liquid phase. The slender "sticks" represent molecules.

How do we quantitatively describe the amount of orientational order present in a liquid crystal? Since the molecules are not fixed, we

CHOLESTERYL MYRISTATE

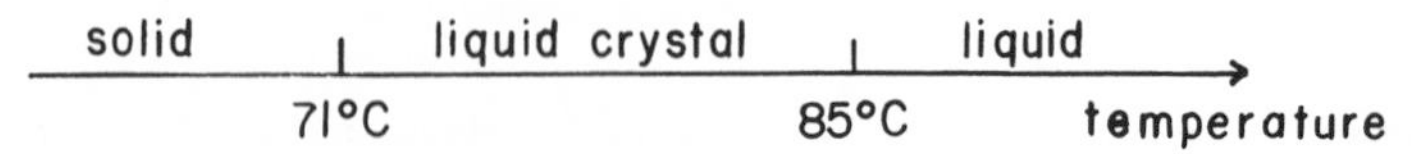

Fig. 1.2 Phase diagram of cholesteryl myristate. A gas phase is not shown because the molecule decomposes at high temperatures.

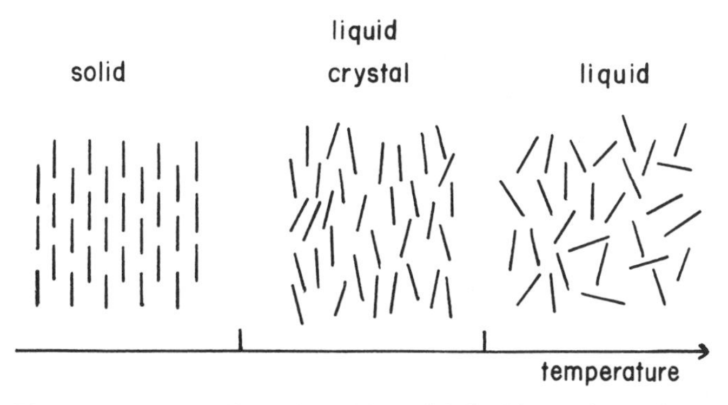

Fig. 1.3 Schematic illustration of the solid, liquid crystal, and liquid phases. The slender ''sticks'' represent molecules.

are forced to describe the order as some sort of average. For example, let us imagine that the direction of preferred orientation in a liquid crystal is toward the top or bottom of the page. This direction can be represented by an arrow, called the *director* of the liquid crystal. There are always two choices as to which way the director points (up or down in this example). The two directions are equivalent in the liquid crystal, so either choice is fine. One way we could visualize performing an average is to take a snapshot of a representative group of molecules at a certain time. Each molecule in our picture is oriented at some angle relative to the director, so our snapshot might look something like figure 1.4. We could measure all the angles and compute the average angle as a measure of the amount of orientational order. The more orientational order present, the closer the average angle would be to zero. A snapshot of the molecules in a liquid with no orientational order would yield values for the angle between 0° and 90°, with angles representing all possible directions distributed at random. We must keep in mind that this random arrangement occurs in three dimensions. Innumerable orientations make an angle of 90° to the director but only one makes an angle of 0°. There will therefore be many more molecules making angles of 90° with the director as compared to 0°. If we average these measurements, the larger number of molecules oriented at larger angles produces a result greater than 45° (57° to be exact). Using this scheme, therefore, no orientational order means an average angle of 57°, with smaller an-

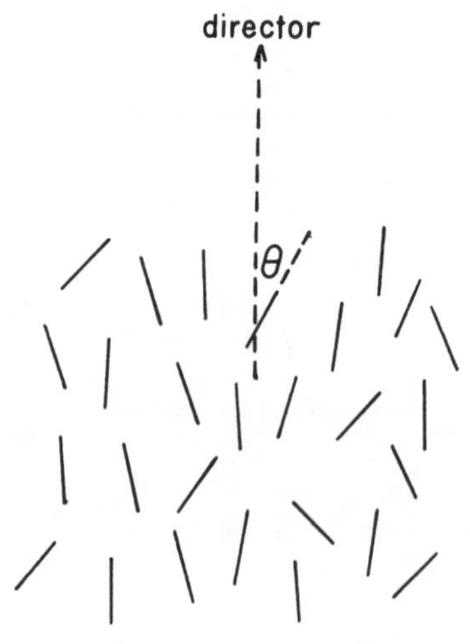

Fig. 1.4 A snapshot of the molecules in the liquid crystal phase. The dashed arrow (director) shows the direction of preferred orientation. Each molecule makes an angle with the director (shown for one molecule as θ).

gles indicating the presence of orientation order. The most order possible (complete alignment) would give an average angle of 0°.

There is nothing wrong with this procedure, but for many reasons a different method is more useful. In this new procedure, the angle the molecule makes with the director is not averaged; instead, the function $(3\cos^2\theta - 1)/2$ is averaged. Since the cosine of 0° is 1, perfect orientational order (all angles equal to 0°) causes this average to equal 1. In addition, in a liquid with no orientational order, the average of this function is 0. This is a more meaningful range of values, and the average of this function is called the *order parameter* of the liquid crystal. It is an extremely important quantity. The order parameter of a liquid crystal decreases as the temperature is increased, and as can be seen in figure 1.5, typical values for the order parameter are between 0.3 and 0.9.

We could have performed this average in another way. Imagine following a single molecule, taking pictures of it at regular intervals. You can imagine measuring the angles in each of these pictures and using them to perform the average. Do you expect that the averages

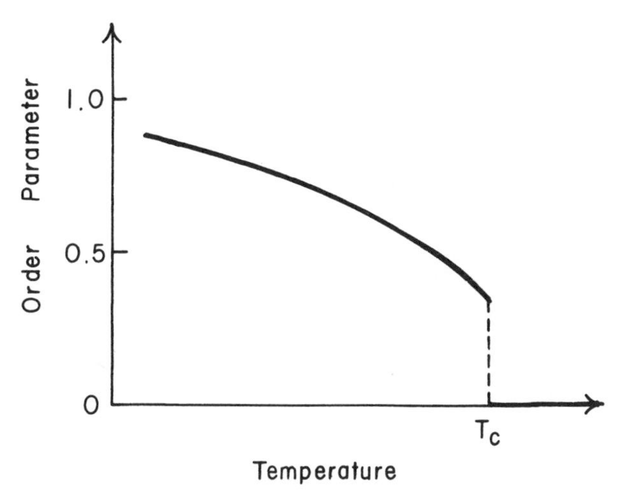

Fig. 1.5 Order parameter variation with temperature in the liquid crystal phase. T_c represents the transition temperature to the liquid phase.

performed in these two different ways would yield different values for the order parameter? The answer is no if you assume that all molecules undergo the same type of random motion. If this assumption is true, the motion one molecule has now is the same as another molecule has at some other time. Performing an average of one molecule's motion at different times should be equivalent to performing an average of many molecules' motion at any one time. Interestingly enough, in all the experiments performed on liquid crystals over the past one hundred years, none of the results have ever contradicted this assumption.

Is a liquid crystal more like a solid or a liquid? This question is important because it turns out that there is a definite answer. When a phase transition takes place from solid to liquid, for example, energy must be supplied to the substance to disrupt the attractive forces that hold the solid together in a highly ordered arrangement. An analogy is the energy you must supply in order to pull two magnets apart. If ice is the substance, you must supply 80 calories of energy for each gram of ice in order to melt it. Likewise, it takes energy to pull the molecules of a liquid away from each other in changing to the gas phase. For example, a gram of water must be supplied with 540 calories of energy before it will all be changed to steam. The amount of energy required to cause a phase transition is called the *latent heat* of the transition and is a useful measure of how different the two phases are. In the case of cholesteryl myristate, the latent heat of the solid to liquid crystal transition is 65 calories/gram, while the latent heat for the liquid crystal to liquid transition is 7 calories/gram. These numbers allow us to answer the question posed earlier. The smallness of the latent heat of the liquid crystal to liquid phase transition is evidence that liquid crystals are more similar to liquids than they are to solids. When a solid melts to a liquid crystal, it loses most of the order it had and retains only a bit more order than a liquid possesses. This small amount of order is then lost at the liquid crystal to liquid phase transition. The fact that liquid crystals are similar to liquids, with only a small amount of additional order, is the key to understanding the many physical properties that make them nature's delicate state of matter.

Although most molecules do not form a liquid crystal phase, it is not a rare occurrence when one does. It is said that an organic chemist indiscriminately synthesizing compounds would find out that about one in every two hundred possesses the liquid crystal phase. After years of experiments, it has become clear what type of molecule is likely to be liquid crystalline at some temperature. First of all, the molecule must be elongated in shape; that is, it must be significantly longer than it is wide. Second, the molecule must have some rigidity in its central region. A molecule that flops around like a piece of cooked spaghetti is unlikely to have a liquid crystal phase. Finally, it seems to be advantageous if the ends of the molecule are somewhat flexible. A good model of a typical liquid crystal molecule is therefore a short pencil with a short piece of cooked spaghetti attached to each end. Why this model works is not difficult to understand. Elongated molecules usually have stronger attractive forces when they are aligned parallel to one another. In addition, elongated molecules bump into each other less when they all tend to point in the same direction, a fact that acts to stabilize aligned phases. The importance of the flexible ends is a bit more subtle. The flexibility seems to allow one molecule to position itself more easily between other molecules as they all chaotically move about.

Although it is possible to observe and identify liquid crystals in bulk samples, one of the most useful and beautiful ways to observe this phase is under a microscope. First, some liquid crystal is placed between two pieces of glass. The thickness of the liquid crystal sample is usually kept small so that light can easily pass through it. The liquid crystal sample is positioned between two polarizers, which are adjusted so that they cross each other. This arrangement would normally ensure that no light comes through to your eye; but, because of the orientational order of liquid crystals, light does reach your eye. (I will discuss polarized light at length later.) Instead, you see a collection of curved lines as shown in plate 1. The lines you see in the microscope are due to defects within the liquid crystal, which also will be discussed later. The name for the liquid crystal phase I have been discussing stems from these defect lines, which reminded some early scientists of threads; therefore, logically, the name *nematic* liquid crystal is derived from the Greek word for thread.

As is often the case in science, things turn out to be more complicated than they first appear. Molecules that possess intermolecular forces and stay parallel to one another form the nematic liquid crystal phase just as I have described. Molecules with intermolecular forces that favor alignment between molecules at a slight angle to one another form a slightly different phase. In this liquid crystal phase, the director is not fixed in space as in a nematic phase; it rotates throughout the sample. A representation of this phase is shown in figure 1.6, again using slender "sticks" to represent the molecules.

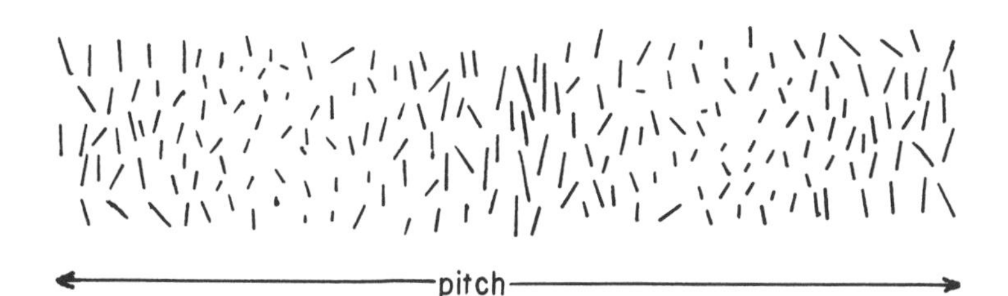

Fig. 1.6 Snapshot of the molecules in the chiral nematic liquid crystal phase. The direction of preferred orientation (director) rotates about a horizontal axis. The pitch is the distance for one full revolution.

The "sticks" appear shorter at some points in the diagram because they are viewed head on.

The best way to visualize the helical pattern formed by the director as it changes its direction is to imagine the motion of a nut as you screw it onto a bolt. As you rotate the nut around and around, the nut moves along the axis of the bolt. This is exactly what the director does in this type of liquid crystal.

Notice that it takes the director a certain distance to rotate one full turn, just as the nut moves a certain distance along the axis of the bolt as it is rotated one complete turn. This distance is called the *pitch* of the liquid crystal. In actuality, the twisted structure repeats itself over a distance equal to one half the pitch, since the director can always be defined pointing in either direction along the preferred direction. This fact will be important when we discuss how light interacts with these twisted liquid crystals.

The most common examples of molecules forming this phase are closely associated with cholesterol, so we call them *cholesteric* liquid crystals. Cholesteryl myristate has a cholesteric liquid crystal phase. The term cholesteric is not a good one, however, since there are many cholesteric liquid crystals that have no connection with cholesterol whatsoever. A more proper name for this phase is *chiral nematic* liquid crystal (chiral simply means twisted). The twist present in chiral nematic liquid crystals produces some spectacular optical properties. I will discuss them later, but some of these properties make chiral nematic liquid crystals appear quite different under a microscope.

Plate 2 shows an example of a chiral nematic liquid crystal under the microscope; notice that defect lines are still present, but the patterns are different due to the twist inherent in the sample. The liquid crystal in plate 2 is oriented with the helical axis along the direction of viewing (perpendicular to the page). This is called the *Grandjean texture*, after the French scientist F. Grandjean, who worked with similar samples around 1920. Plate 3 shows another picture of a chiral nematic liquid crystal, but the helical axes in this sample are perpendicular to the viewing direction. The pitch of this liquid crystal is long enough to be visible under the microscope, so all the lines in the picture are points where the director rotates back to the same position. When this occurs, it is called the *fingerprint texture*.

A substance may possess either the nematic liquid crystal phase or the chiral nematic liquid crystal phase, but not both. However, there is another type of liquid crystal phase that can occur as the only liquid crystal phase a substance possesses or at a temperature below the nematic or chiral nematic phase of a substance. This third liquid crystal phase is called the *smectic* phase, from the Greek word for soap. The early investigators noticed that these liquid crystal phases possessed mechanical properties reminiscent of soaps. In fact, the thick "goo" usually found in the bottom of a soap dish is a liquid crystal phase (I will discuss this type later) and has all the properties of a smectic liquid crystal. In the smectic phase, not only is the small amount of orientational order of liquid crystals present, but there is also a small amount of positional order. Again, the molecules are free to bounce around quite randomly, but in this phase they tend to point along the director and arrange themselves in layers. To be more exact, a snapshot in time would reveal that slightly more molecules tend to be positioned in regularly spaced planes with fewer molecules in between. Likewise, following a single molecule would reveal that it spent slightly more time in these planes than between them. Figure 1.7 illustrates this small amount of both orientational and positional order. Notice that there are two slightly different smectic phases. In the smectic *A* phase the director is perpendicular to the planes, while in the smectic *C* phase the director makes an angle other than 90° to the planes. Smectic liquid crystals have a unique appearance under the microscope; an example can be found in plate 4.

In both the smectic *A* and smectic *C* phases, the molecules randomly diffuse within each plane. No positional order exists within each plane, so in a sense the positional order is in one dimension only. However, other smectic liquid crystal phases exist in which the molecules are somewhat ordered within each plane. In other words, a molecule diffusing through the plane spends more time at special locations than at other locations. The positional order is three-dimensional now. Various arrangements of these special locations are possible, and these phases have been given names like smectic *B* or smectic *E*. The designation of smectic phases by letters is more historical than physical, in that additional letters have been used each time a new smectic phase is discovered. A recent list of these smectic phases included the letters *A* through *K*. The additional order of these

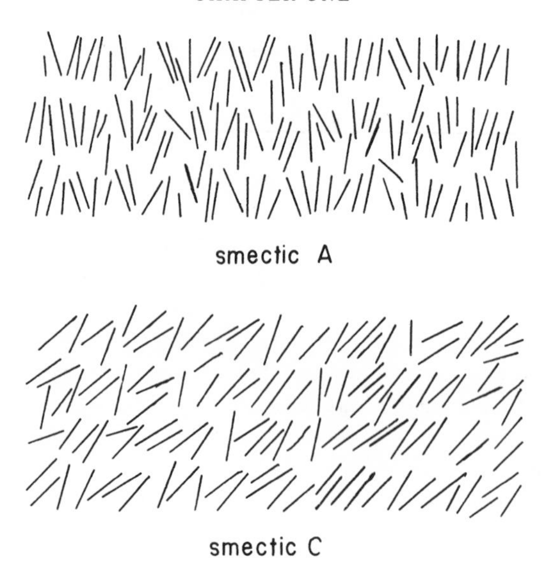

Fig. 1.7 Snapshot of the molecules in two types of smectic liquid crystal phases. The layers are perpendicular to the director in the smectic *A* phase, but make an angle other than 90° in the smectic *C* phase.

smectic phases makes them appear quite different under a microscope. Plate 5 is a picture of a smectic *B* phase.

Although many compounds exhibit only one liquid crystal phase, it is not unusual for a single substance to possess more than one. A compound called *p*-azoxyanisole (PAA) has only a nematic phase; its phases are shown in figure 1.8. The figure also contains a schematic of the PAA molecule. For simplicity, some of the carbon atoms are denoted by points where one or more lines (representing bonds) meet. The hydrogen atoms bound to these carbon atoms are also omitted. You can tell where they are by realizing that every carbon atom must share its four electrons. The number of hydrogen atoms bound to a carbon atom is therefore four minus the number of lines (other bonds) coming together to represent the carbon atom.

A compound slightly different from PAA is called 4-*n*-pentylbenzenethio-4′-*n*-decyloxybenzoate (abbreviated as $\overline{10}$S5). It possesses three liquid crystal phases, as shown in figure 1.9. Cholesteryl my-

PAA

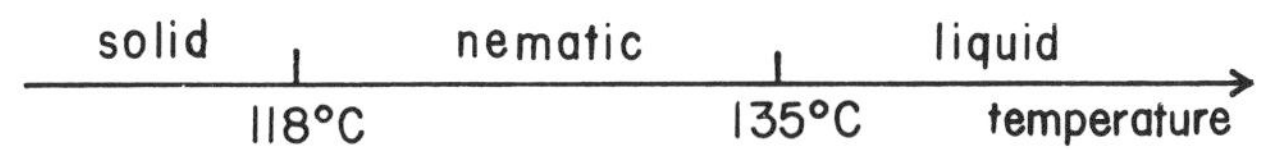

Fig. 1.8 Phase diagram for *p*-azoxyanisole (PAA). In the diagram of the molecule, intersecting lines (which represent chemical bonds) denote carbon atoms. For simplicity, the hydrogen atoms bonded to these carbon atoms have been omitted.

$\overline{10}$S5

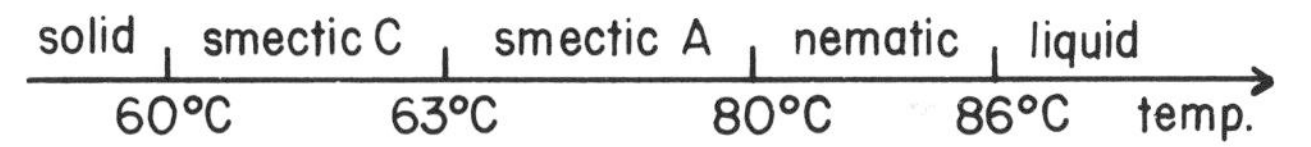

Fig. 1.9 Phase diagram and molecular structure of 4-*n*-pentylbenzenethio-4′-*n*-decyloxybenzoate ($\overline{10}$S5).

ristate has both a chiral nematic and a smectic A phase, so its full diagram is as follows (see figure 1.10).

Discotic Liquid Crystals

So far the entire discussion has been devoted to the liquid crystal phases formed by rodlike molecules. The reason for this is simple; these phases are the most common liquid crystal phases and therefore the most well known. In 1977, however, researchers in India discovered that disklike molecules also form liquid crystal phases in which

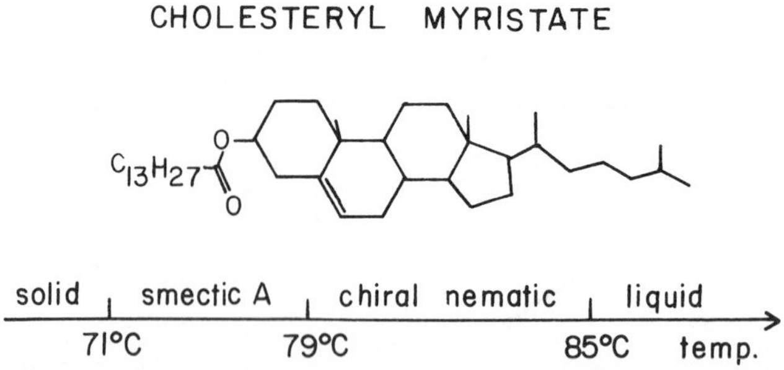

CHOLESTERYL MYRISTATE

$C_{13}H_{27}$

solid | smectic A | chiral nematic | liquid

71°C 79°C 85°C temp.

Fig. 1.10 Complete phase diagram for cholesteryl myristate along with its molecular structure.

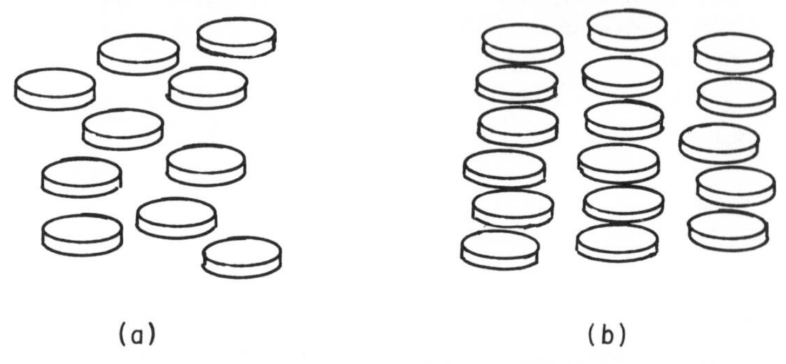

Fig. 1.11 Illustration of the (a) nematic and (b) columnar discotic liquid crystal phases. The flat disks represent molecules.

the axis perpendicular to the plane of the molecule tends to orient along a specific direction. These phases and the molecules that form them are called *discotic* liquid crystals.

The most simple discotic phase is also called the *nematic* phase, because there is orientational order but no positional order. Such a phase is shown in figure 1.11(a), where the molecules have been drawn as flat disks. The molecules move about quite randomly, but on average the axis perpendicular to the plane of each molecule tends to orient along a special direction called the director. As evident from the figure, this phase looks like a pile of coins. The *columnar* or *smectic* discotic phase is shown in figure 1.11(b). In addition to the orientational order present in the nematic discotic phase, most of the molecules tend to position themselves in columns. The columns are arranged in a hexagonal lattice. A nice analogy to this phase is a set of stacked coins, with the stacks positioned in the centers and vertices of connected hexagons. However, one must be careful about such analogies. The coins in a stack have a great deal of positional order (i.e., the distance between coins is fixed and the same for all coins), whereas the molecules of a columnar or smectic discotic liquid crystal are positioned in the stack quite randomly. The positional order in this phase is therefore two-dimensional. *Chiral nematic* discotic liquid crystals also exist. In this phase the director rotates in a helical fashion throughout the sample, just as in the case of rodlike chiral nematic liquid crystals. An example of a molecule possessing discotic liquid crystal phases is shown in figure 1.12 along with a temperature diagram of these phases. Notice that this molecule has a fairly rigid, planar center with hydrocarbon chains emanating in all directions. These features are common to just about all discotic liquid crystal molecules.

Although the structure of discotic liquid crystals is quite different

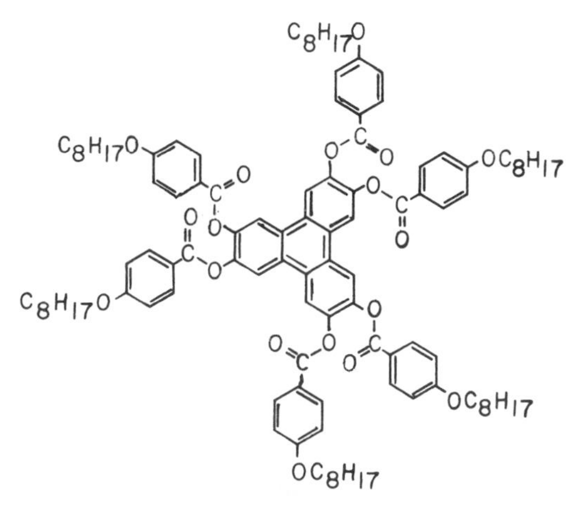

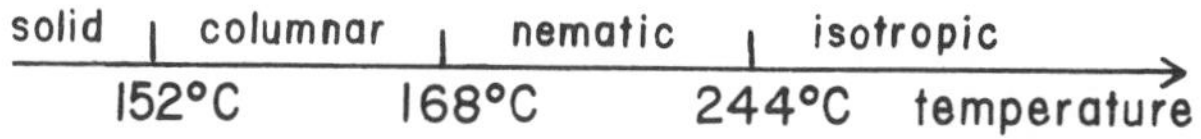

Fig. 1.12 Phase diagram and molecular structure of a typical discotic liquid crystal.

from other types of liquid crystals, their appearance under the microscope is similar. Plate 6 shows a "crystal" of a discotic liquid crystal as it forms from the isotropic liquid.

Other Types of Liquid Crystals

One area of modern technology that also deals with liquid crystal phases is the polymer industry. *Polymers* are extremely long and slender molecules that form when chemical reactions link shorter molecules together. A bowl of cooked spaghetti is an especially good model for most polymers. Two types of polymers give rise to liquid crystal phases. The first is composed of fairly rigid segments (like the central part of a liquid crystal molecule) connected together end-to-end by flexible segments (like the ends of a liquid crystal molecule). Although these long polymers move around and collide with each other in the liquid crystal phase, the rigid segments tend to remain pointing in one direction. The second type of polymer is one composed of a single, completely flexible polymer with rigid segments attached along its length by short flexible segments. In the liquid crystal phase of this type of polymer, the long flexible part winds its way throughout the substance without any orientational or positional order; but the rigid segments attached to it exhibit the orientational order typical of liquid crystal molecules. Figure 1.13 illustrates both of these types of liquid crystal polymers. Nematic, chiral nematic, and smectic phases have been found in polymers, with some possessing more than one liquid crystal phase. Their appearance under a microscope resembles other liquid crystals in many ways. A photograph of a nematic polymer liquid crystal is shown in plate 7. Liquid crystal polymers are an exciting new field of modern technology, which is covered in some depth in chapter 9.

Before our introduction to the liquid crystal phase can be complete, we must consider one other class of compounds where liquid crystal phases are important. In the discussion thus far, I have considered only pure substances, and temperature changes have caused the liquid crystal phases to come and go. This class of liquid crystal substances is known as *thermotropic* liquid crystals. In some cases when two different substances are mixed together, the mixture can exhibit different phases not only as the temperature is changed, but also as

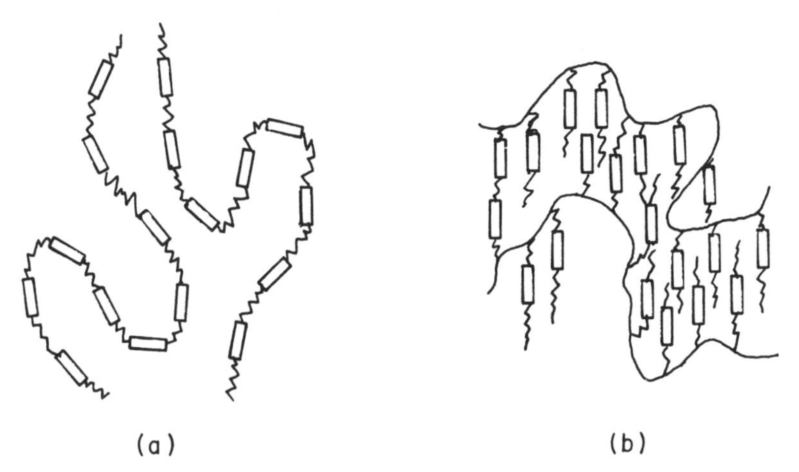

Fig. 1.13 Illustration of the orientational order in (a) a main chain and (b) a side chain liquid crystal polymer. The rectangles represent rigid segments and the zig-zag lines represent flexible segments of the polymer.

the concentration of one component of the mixture is varied. When the liquid crystal phase is dependent on the concentration of one component in another, it is called a *lyotropic* liquid crystal.

The easiest way to put together a lyotropic liquid crystal mixture is to start with a molecule that has end groups with different properties. For example, one end of the molecule could show an affinity for water, while the other end tends to exclude water. When such molecules are placed in water, the ends that exclude water tend to arrange themselves together, allowing the other end of the molecules to be in contact with the water. This effect results in structures of various shapes (spheres and cylinders are common), which themselves can be positioned in a very specific arrangement. Again the molecules are free to roam about. But as they do so, they retain the orientational and positional order of these structures, and are therefore proper liquid crystalline phases. Molecules of this type can be represented by a circle (water seeking) connected to a tail (water excluding). Two of the many structures these molecules form are shown in figure 1.14.

There are two important examples of molecules that behave as I have been describing. The first are *soap* molecules, and the ability of

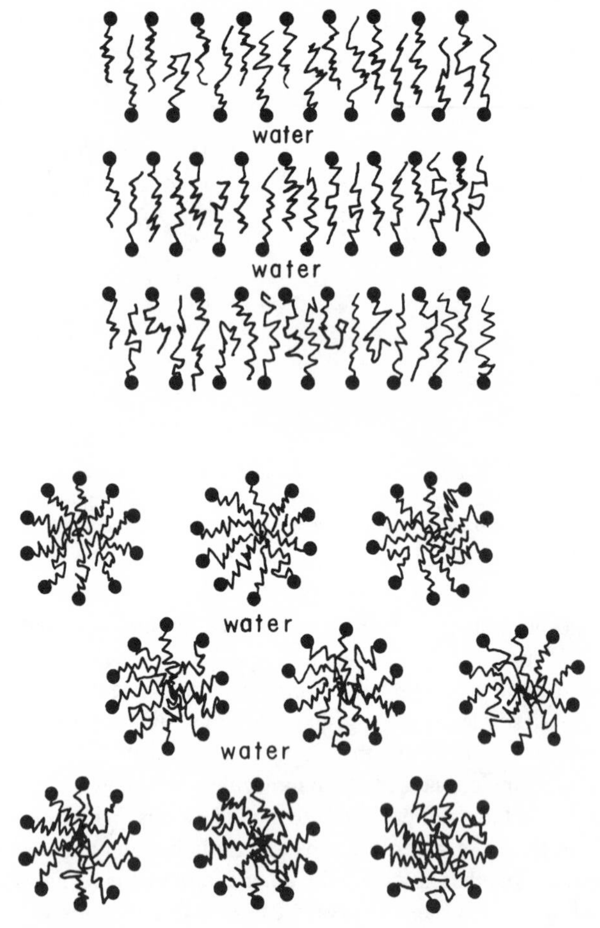

Fig. 1.14 Cross-sections of the lamellar (top) and hexagonal (bottom) lyotropic liquid crystal phases. The round end of each molecule represents a hydrophilic (water-seeking) group, while the zig-zag line depicts a hydrophobic (water-fearing) group.

soap to dissolve oil and dirt is directly related to the ordered structures the soap molecules form in water. The second example are biologically important molecules called *phospholipids*. The ordered structures these molecules form are found everywhere in biological systems and include the cell membrane itself. In both of these types of molecules, one end of the molecule is charged and tends to associate with water, while the other end is composed mostly of carbon and hydrogen that tend to exclude water. Lyotropic liquid crystals are important not only for their biological significance, but also because of their use as surfactants. Chapter 8 addresses both of these areas.

CHAPTER 2

The Story of Liquid Crystals

For one week during the summer of 1988, over seven hundred scientists from thirty-one countries gathered in the city of Freiburg in the Federal Republic of Germany for the twelfth International Liquid Crystal Conference. The number of researchers participating in this conference testified to the importance of liquid crystal research and technology in today's world. One special aspect of this conference was that it celebrated the one hundredth anniversary of the ''discovery'' of liquid crystals. The story of these hundred years, how interest in liquid crystals waxed and waned, and how the field blossomed twenty years ago, is truly a fascinating one.

Much of what we know about the early scientific work on liquid crystals is due to Dr. H. Kelker from the University of Frankfurt. Part of the one hundredth anniversary celebration at the Freiburg conference was a lecture on the ''discovery'' of liquid crystals by Dr. Kelker.

BEFORE THE ''DISCOVERY''

Although individual scientists are frequently given credit for important scientific discoveries, in many cases there are other scientists who also deserve a great deal of credit. Usually these people perform experiments or introduce new ideas that are very similar to those of the ''discoverer.'' Yet these workers for one reason or another do not put their results or ideas together in a way that fully captures the significance of their new work. Sometimes the ''discoverer'' is the scientist who recognizes the meaning of this new work, performs a few experiments that show the true meaning in an unambiguous way, and then clearly explains it to the rest of the world. It is therefore not surprising to hear comments like, ''I was in the right place at the right time,'' coming from these ''discoverers.''

The story of liquid crystals starts with a few European investigators who observed some new and interesting phenomena, but who never fully realized exactly what was happening in their experiments. These early scientists are not given the credit for discovering liquid crystals although we are now sure they observed liquid crystal phases in their laboratory. In fact, their names are often omitted in historical accounts.

The time frame of this early work is 1850 to 1888, and three different types of experiments were involved. The first type concerned the study of some biological specimens using a microscope. It was noticed that the outer covering of a nerve fiber formed soft and flowing forms when left in water. It was also observed that these forms produced unusual effects when polarized light was used. One European scientist, Rudolf Virchow, left a report of his experiments in his archives; the German ophthalmologist C. Mettenheimer reported his observations in a scientific publication; and a short monograph by G. Valentin that summarized the results of polarization microscopy on biological materials was published in Leipzig. It is easy to understand why these scientists found their results interesting; liquids produce no unusual effects when polarized light is used, whereas solids do. The biological specimens were not solid, and yet the unusual polarization effects were present. Further progress using these specimens was probably hindered by the fact that the biological specimens were complicated structures, thus making analysis of the phenomenon quite difficult.

The second type of early experiment involved the study of how substances crystallize. In Karlsruhe, the German physicist Otto Lehmann constructed a heating stage for his microscope, allowing him to observe how materials crystallized as he slowly lowered the temperature of his samples. Although at this time biologists and physicians were using polarized light to view biological samples, Lehmann was one of the first physical scientists to use a polarizing microscope when he later added polarizers to his heating stage microscope. This instrument was a significant technical advancement that later allowed Lehmann to be one of the most important early contributors to the field of liquid crystals. A sketch of his apparatus appears in figure 2.1. The success of Lehmann's microscope in furthering scientific knowledge is an excellent example of how a technological achievement can dramatically affect the course of science. With this

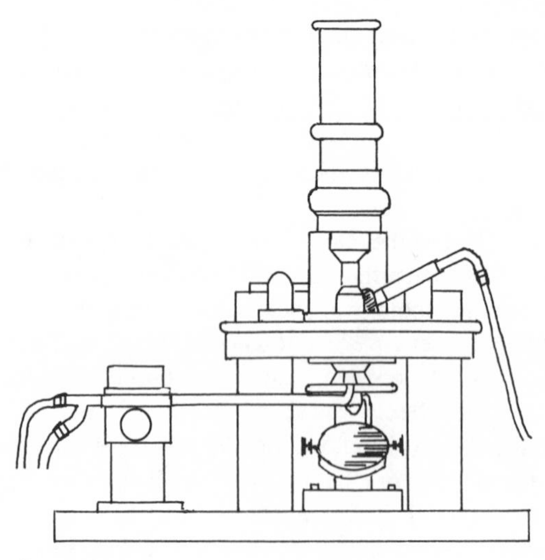

Fig. 2.1 Otto Lehmann's microscope and heating stage. A flame from the burner (under the stage) heated the stage while air from the tube (above the stage) cooled it. Careful adjustment of these two devices allowed Lehmann to precisely control the temperature of his samples.

new apparatus, scientists could control the environment of their samples and observe their behavior under a microscope. This instrument allowed them to perform experiments that were previously impossible and to observe phenomena that earlier were hidden from the investigators' eyes. Liquid crystal research would never be the same. In fact, a polarizing microscope with a precision heating stage is a standard piece of modern equipment; today no liquid crystal laboratory works without one.

Lehmann noticed that some substances did not crystallize from a clear liquid but changed instead to an amorphous form, which then crystallized. He did not realize at this point that a new phase of matter was being observed; he conjectured that it was somehow related to the transition from the liquid phase to the solid phase. It is easy to understand why Lehmann might have thought in this way. Phase transitions do not always occur in the idealized way. For example, the presence of impurities causes phase transitions to occur over a narrow temperature range where both phases coexist. Certainly Lehmann had observed many phase transitions under the microscope and knew they varied significantly from one substance to the next. The amorphous form he saw in his microscope might have struck him as just another less than ideal solid to liquid phase transition.

The third type of experiment performed in this time period was one with compounds synthesized from cholesterol. P. Planer in the city of Lvov (now part of the Soviet Union), the German chemist W. Lobisch, and B. Raymann in Paris all reported that these compounds displayed striking colors when cooled. Although it certainly was an unusual phenomenon, none of these people had any idea that these colors were coming from a phase of matter other than the solid or liquid phase.

One other investigation from this time period deserves to be mentioned. Around 1850, the chemist W. Heintz, who was studying natural fats, reported that stearin possesses an unusual melting behavior. It turns cloudy around 52°C, is completely opaque by 58°C, and becomes clear at 62.5°C. He and a colleague at one point referred to the change at 62.5°C as a second melting point. As will be evident in the next section, these observations were extremely similar to the experiments that forty years later led directly to the "discovery" of liquid crystals.

The "Discovery"

The person generally given credit for "discovering" liquid crystals is an Austrian botanist named Friedrich Reinitzer. Reinitzer's main interest was the function of cholesterol in plants. Of course, the structure of cholesterol was not known at the time. In 1888 when Reinitzer observed the melting behavior of an organic substance related to cholesterol, he also described it by saying that it possessed two melting points. At 145.5°C it melted to a cloudy liquid, and at 178.5°C this cloudy liquid turned into a clear liquid. He also described some of the same color phenomena that had been reported by earlier scientists working with cholesterol derivatives: a blue color briefly appears upon cooling when the clear liquid turns cloudy and a blue-violet color occurs just before the cloudy liquid crystallizes.

Why Reinitzer stated that the substance seemed to have two melting points, while earlier scientists frequently associated unusual behavior of this sort with the crystallization process, is one of those intriguing questions that often arise in the history of science. Was he aware of the work of Heintz? We cannot be certain. What we do know is that twenty years later he mentions Heintz's experiments in a letter to Otto Lehmann. Regardless of whether or not Reinitzer was the first to describe a substance as having two melting points, his description of the cholesterol derivative opened a line of inquiry that turned out to be extremely fruitful. If the solid phase melted at 145.5°C, what was melting at 178.5°C? Might this be a form of matter other than the solid or liquid phase? Because he described what he saw in a way that posed the right questions for future investigators, Reinitzer is usually called the ''discoverer'' of liquid crystals.

The substance Reinitzer observed was in fact cholesteryl benzoate, a chiral nematic liquid crystal that has been the object of many investigations over the years. Reinitzer made this compound from cholesterol, a natural substance found in both plants and animals. Interestingly enough, the blue color Reinitzer saw just as the sample turned from clear to cloudy became a very important problem for condensed matter physicists around 1980; but a discussion of this must wait until a later chapter.

Reinitzer knew of the work of Lehmann and saw the connection between his observations and Lehmann's findings. He therefore sent some of these samples to Otto Lehmann, who at this time was a professor of natural philosophy (physics) in Germany. Lehmann performed many experiments on this substance with his heating stage microscope. In one report he includes a drawing that is extremely similar to the picture of a chiral nematic shown in plate 2. Lehmann described Reinitzer's substance along with substances he received from other investigators in a variety of ways during this time period. First he called them soft crystals that were almost fluid, then the term floating crystals was used, which was later replaced with crystalline fluids. He gradually became convinced that the cloudy liquid was a uniform fluid phase, but one that affected polarized light in a manner typical of solid crystals, not liquids. This combination of characteristics—flow properties like a liquid and optical properties like a solid—finally led Lehmann to label these substances liquid crystals. In spite of many subsequent arguments over nomenclature, this was the name that eventually survived.

I must make note of the symbolism in Reinitzer's sending of his samples to Lehmann. Reinitzer was an Austrian botanist and Lehmann was a German physicist, so the act represented a collaboration between scientists representing both two scientific disciplines and two different countries. Thus, from the beginning, liquid crystal research was an international and multidisciplinary field, a fact that has grown more true with each passing year. Chemists, physicists, biologists, engineers, and medical doctors from almost all of the industrialized countries of the world are involved in liquid crystal research. They attend the same scientific meetings and spend time as visiting scientists at each other's laboratories. The spirit of international and interdisciplinary collaboration displayed by these two early liquid crystal researchers is still the hallmark of the field.

After the "Discovery"

Lehmann remained the dominant figure in liquid crystal research around the turn of the century. He experimented with the first nematic liquid crystal, which was synthesized by two German chemists, Ludwig Gattermann and A. Ritschke. This compound was also the first liquid crystal not based on a natural substance. In fact, Gattermann and Ritschke supplied him with a series of substances, and Lehmann noticed that some liquid crystals behaved differently from others. He used different words to describe them but did not choose to call them two different types of liquid crystals (now known as nematic and smectic). Lehmann also observed that a solid surface in contact with a liquid crystalline substance causes the liquid crystal to orient in a certain direction. This idea was of paramount importance when modern scientists began to experiment with liquid crystal displays.

Not all of Lehmann's ideas were accepted by other investigators. Gustav Tammann, a solid state chemist, and Walter Nerst, a physicist, argued that the unusual phenomena could be explained by the fact that the substance was a mixture or emulsion of two distinct compounds or phases. Lehmann used the results of his own experiments, together with those of the physical chemist Rudolf Schenk, to argue more and more convincingly that a single phase was involved.

Another important contributor at this time was the German chemist Daniel Vorlander, who worked in Halle (now in the German Democratic Republic). Workers in his institute synthesized many new liquid crystalline substances and were the first to observe a single substance that possessed more than one liquid crystal phase. Out of this work Vorlander was able to identify what kinds of substances were likely to be liquid crystalline. Being able to make this identification was of obvious practical importance, but his suggestion that a linear molecular shape was important influenced both theoretical and experimental work for many years to come. As time went on, more results about new liquid crystals poured out of Vorlander's group, with over eighty doctoral theses being written on liquid crystals between 1901 and 1934. Today the Martin Luther University in Halle still boasts an important liquid crystal institute. In fact, there has been continuous liquid crystal research at Halle from about 1900 to the present.

Vorlander's suggestions as to what kinds of substances were likely to have liquid crystalline properties were a strong influence on the theoretical and experimental work of the physicist Emil Bose. Bose attempted to construct a complete theory of liquid crystals based on molecular structure. This was an important contribution, because up to this point there had been considerable confusion on Lehmann's part as to what the ordering unit was and what changes (chemical and/or physical) took place at the transition. Bose's theoretical work also represented a strong argument against the hypothesis that liquid crystals result from some sort of emulsion. A little less than ten years later, Max Born suggested that the basic interaction between the linear molecules that causes the liquid crystal phase is due to a slight separation of positive and negative charges on the molecules.

Lehmann himself was partially responsible for introducing the study of liquid crystals in France. In 1909 he was invited to a conference in Paris. The culmination of the subsequent work in France was the publication in 1922 of a paper by Georges Freidel describing the different liquid crystal phases. In this paper Freidel proposed the classification scheme using the words nematic, smectic, and cholesteric. The accepted description of the liquid crystal phase now involved the idea of molecular ordering, laying to rest numerous other suggestions as to the origin of liquid crystallinity. He also explained that the lines one sees in liquid crystal phases under the microscope were defect structures, which represent drastic changes in the direction of orientation. Through careful analysis of these defect structures, Freidel correctly deduced that smectic liquid crystals possess a layered structure. Additionally, he clearly understood that a liquid crystal could be oriented by an electric field. The effect of electric and magnetic fields on liquid crystals later became the subject of great attention.

The period between 1922 and World War II was characterized by advancement in several different areas by scientists in many countries. Theoretical work into the elastic properties of liquid crystals grew out of the work of Carl Oseen in Sweden and reached fruition in the "continuum theory" of F. C. Frank in England. Using these new theoretical ideas, these scientists explained why liquid crystals adopted various orientational configurations. X-ray experiments in France and Germany revealed in a most unambiguous way that liquid crystals possess more order than liquids, but less than solids. The action of electric and magnetic fields on liquid crystals began to be understood, mainly due to experiments performed in Holland and the Soviet Union. Finally, some progress was made in understanding the light scattering properties of liquid crystals, explaining for the first time why liquid crystals appear cloudy.

Progress was slow during the war years, but two new areas of success should be pointed out. First, the flow properties of liquid crystals in high electric and magnetic fields was investigated. Second, the "degree of order" was described by the order parameter

$$S = \text{average of } (3\cos^2\theta - 1)/2.$$

Choosing to use this order parameter may seem like a fairly unimportant step, but the opposite turns out to be true. We now know from a combination of both theoretical and experimental results that a full description of the orientational order present in liquid crystals is quite a complicated task. Fortunately, the above order parameter comes closer to describing the actual orientational order than any other single parameter. The scientists of the 1940s suspected this result to be the case, and nearly all future work was based on descriptions utilizing this order parameter.

Recent Developments

Interest in liquid crystals all but disappeared after World War II. In describing a possible reason for this lack of interest, F. C. Frank remarked in 1958 that too many people felt that all the important problems concerning liquid crystals had been solved. Another reason given is that writers of chemistry and physics textbooks during this period did not discuss the liquid crystal phase, so many scientists did not know that certain unusual melting phenomena involved a completely different phase of matter. James Fergason, one of the pioneers in the development of liquid crystal displays, felt that the most important force behind this period of quiescence was the apparent lack of any practical applications for liquid crystals.

Whatever the reason for the lack of interest immediately after World War II, the situation began to change shortly before 1960. At this time a few individuals undertook a general reexamination of liquid crystals in the hope of learning more about their molecular structure, optical properties, and technical possibilities.

In 1957 Glenn Brown, an American chemist, published an article reviewing the liquid crystal phase. In 1958 the Faraday Society of London held a conference on liquid crystals. At about the same time, scientists at the Westinghouse Research Laboratories began a research project focused on liquid crystals. A group led by I. G. Chistiakoff began work in the Soviet Union; and in 1962 George Gray, an English chemist, published a full-length book describing the molecular structure and properties of liquid crystals. Brown was instrumental in founding the Liquid Crystal Institute at Kent State University, and also organized the first of a series of international liquid crystal conferences that continue to this day. The conference in Freiburg was part of this series.

Immediate progress was made in understanding the molecular structure of liquid crystal forming compounds. Two German physicists, Wilhelm Maier and Alfred Saupe formulated a microscopic theory for the liquid crystal phase that did not hypothesize a separation of charge on the molecules. The importance of this last result should be emphasized. For the first time, liquid crystal researchers possessed a theory that predicted the behavior of the phase starting with realistic characteristics of the individual molecules. Also, it quickly became apparent that liquid crystalline substances had the ability to detect extremely small changes in temperature, mechanical stress, electromagnetic radiation, and chemical environment; the door was thus opened to a number of possible applications. In 1968, two scientists at RCA demonstrated that a thin layer of liquid crystal was capable of switching from cloudy to clear when an electrical voltage was applied. Although this was the first liquid crystal display (LCD), it required too high a voltage, consumed too much power, and produced a display of poor quality. Three years later, two scientists working in Switzerland and one in the United States reported that a liquid crystal cell could be turned from clear to black by an electrical voltage. The improved quality of this display together with its low power consumption made this device worthy of application. Within ten years liquid crystal displays that required extremely little power were being used by manufacturers of battery operated equipment such as watches and calculators.

Perhaps the most important advance during this period was the synthesis of the first moderately stable room temperature liquid crystal, *p*-methoxybenzylidene-*p*-*n*-butylaniline (MBBA). The phases of MBBA along with its molecular structure are shown in figure 2.2.

The combination of both basic scientific progress and new technological ideas caused an explosion in the number of researchers in the field. In the area of LCDs alone, at least six different types of displays were developed, including some that used chiral nematic and smectic liquid crystals. Research papers began to fill volumes each year, and new applications appeared regularly. Research conferences on liquid crystals grew in size from roughly fifty participants in the early 1960s to nearly seven hundred in the late 1980s. Groups

MBBA

$CH_3-O-C_6H_4-CH=N-C_6H_4-C_4H_9$

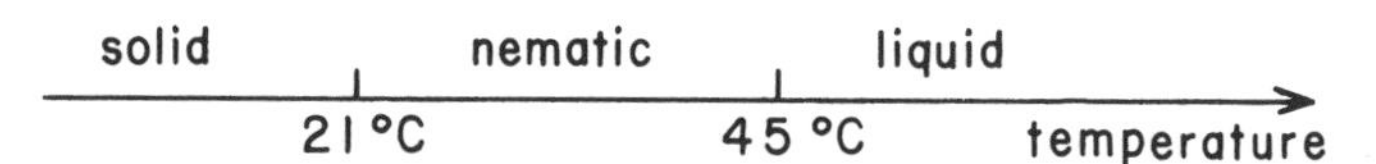

Fig. 2.2 Phase diagram and molecular structure of *p*-methoxybenzylidene-*p*-*n*-butylaniline (MBBA), the first fairly stable room temperature liquid crystal.

of physicists, chemists, and engineers studying liquid crystals formed at major universities and industrial laboratories throughout the world. This period was clearly the most exciting time in the history of liquid crystal research, and much of our detailed knowledge about liquid crystals stems from work done at this time.

The study of liquid crystals has played an important role in our increasing understanding of how molecules behave cooperatively and how molecular structure influences this behavior. Chemists synthesizing new liquid crystal compounds have contributed to our understanding of organic synthesis in general. Liquid crystals have been used as a solvent or medium in which to probe other substances. This flurry of activity has also produced liquid crystal displays for watches, calculators, clocks, telephones, cameras, office equipment, personal computers, miniature TVs, automobile dashboards, and windows that can change from clear to opaque. Other applications include liquid crystal thermometers and temperature-sensing films, high-strength liquid crystal polymers, and surfactants for the oil recovery industry. Progress in our understanding of the liquid crystal phase has also aided our understanding of the cell membrane and of certain diseases, such as sickle-cell anemia and arteriosclerosis. Many of these important contributions will be discussed fully in later chapters, but we can be certain that just as many new developments will be made by the scientists and engineers of the future.

CHAPTER 3

Electric and Magnetic Field Effects

The response of liquid crystals to electric and magnetic fields is perhaps the best illustration of the delicate nature of this phase of matter. Solids, liquids, and gases respond to electric and magnetic fields, but the response is minimal even when strong fields are applied. Liquid crystals, on the contrary, respond to even weak electric and magnetic fields with significant structural changes. This chapter explores these effects by first describing the reasons behind this delicate response to electric and magnetic fields and then discussing the many different ways in which this response can take place.

ANISOTROPY

Although liquid crystals are fluids, the fact that the molecules on average orient themselves along a certain direction in space has a profound effect on the properties of the phase. The best way to understand this is to consider the case of liquids first. In the liquid phase (or the gas phase) the completely disordered motion of the molecules produces a phase in which all directions in space are equivalent. In other words, any property measured along one direction of space has the same value when measured along any other direction. For example, suppose sound is made to propagate through a liquid. If we measure how the sound loses its intensity as it passes through the liquid, we obtain the same answer no matter in which direction the sound is traveling. The property of obtaining the same result regardless of direction is called *isotropy* and a phase that has this property is called an *isotropic* phase. All liquids and gases are isotropic.

If one imagines our picture of the liquid crystal phase, it is not difficult to realize that this phase is not isotropic. At every point in a liquid crystal the molecules define a special direction by spending more time pointing along this direction than any other direction. Sound traveling along the director of a liquid crystal encounters molecules that tend to lie along the direction of propagation, while sound traveling in other directions encounters molecules oriented on average at an angle to the direction of propagation. Since a group of molecules responds differently to sound according to the direction the sound is traveling relative to the long axis of the molecules, sound traveling along the director will lose its intensity in a different way than sound traveling in any other direction. How the sound loses its intensity in these two cases depends on the actual molecule that makes up the liquid crystal. Typically, sound loses its intensity over a shorter distance when traveling along the director. This property is called *anisotropy* (no isotropy) and accordingly the liquid crystal phase is called an *anisotropic* (not isotropic) phase.

For most people, the difference between a solid and a liquid is obvious. Solids are "rigid" and liquids flow. As simple as this distinction is, there are some substances that are not easily placed in one category or the other. For example, some plastics flow so slowly that an observation over a short time interval might lead to the conclusion than it does not flow. Classifying it as a solid, however, would be misleading. Anisotropy is another property used to distinguish solids from liquids. Since no liquid is anisotropic, any anisotropic substance must be a solid (unless it is a liquid crystal). The correspondence between anisotropy and solids is not perfect, however, because some solids are isotropic. By using several criteria to classify whether a substance is a liquid or a solid, scientists can deal more successfully with those substances that do not behave in the normal manner. Anisotropy is the most important characteristic shared by both liquid crystals and solids. Otto Lehmann recognized this fact when he noted that light interacted with this "new substance" just as it did with anisotropic solids. No liquid was capable of such behavior.

This anisotropy manifests itself in many physical properties. When external fields (electric or magnetic) are applied to a liquid crystal, the phase responds differently depending on whether the field is applied along the director or at an angle to it. Since light is an electromagnetic wave, how light propagates through a liquid crystal depends on its direction relative to the director. A very different example concerns the mechanical properties of liquid crystals. Imagine placing some liquid crystal between two flat plates and measuring the force necessary to slide one plate over the other. As shown in figure 3.1, the director of the liquid crystal can point in three different directions relative to the plates and direction of movement. The force necessary to slide the plates over each other is different for each case, being nearly the same for cases (a) and (b) but much higher for case (c). No such effect exists for a liquid placed between the two plates. These results are not difficult to reconcile if one realizes that this situation is similar to what happens when logs float down a narrow river. If the logs point downstream, they flow without the problems encountered if they flow pointing across the river. These two cases are analogous to (b) and (c) in the liquid crystal, respectively. The analogy to case (a) in the liquid crystal is not possible, since logs floating in water rarely orient themselves vertically!

The Electric Field

Electricity is the phenomenon associated with charged objects. As we all know, the charge on such objects comes in two varieties (positive and negative), with opposite charges attracting each other and like charges repelling each other. The two charged objects need not be in contact to experience this force; two charged objects separated by any distance attract or repel each other, although the strength of the force does decrease as the distance between the two objects in-

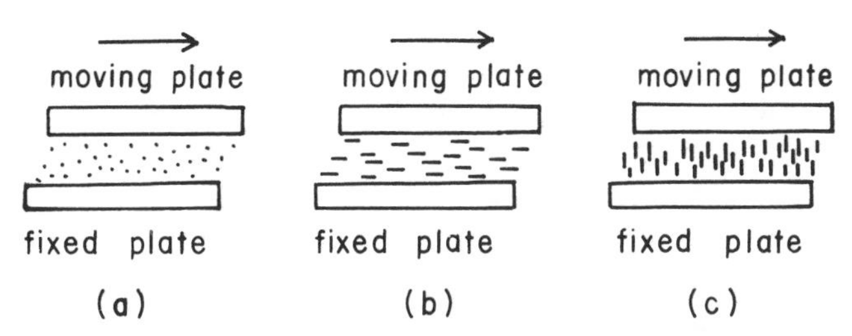

Fig. 3.1 Testing anisotropy in a liquid crystal. The force necessary to slide the top plate over the bottom plate with liquid crystal in between depends on the orientation of the director. The necessary force is less with the director pointing out of the page (a) or along the direction of plate movement (b) than when the director points up or down (c).

creases. Scientists use the idea of an *electric field* to describe the fact that charged objects experience a force due to other charged objects even though there may be some distance between them.

The idea of an electric field is a simple one. Consider a single charged object sitting at a point in space. If a second charged object is placed near it, an electric force between the two objects occurs. The presence of the first charged object somehow alters the space around it, since the second charged object behaves differently if placed in this space depending on whether the first object is present or not. To describe the fact that the space is altered, we say that the first charged object produces an electric field around it. When a second charged object is placed in this electric field, a force is produced. In our minds, therefore, we have divided this phenomenon into two stages: the first charged object produces an electric field, which in turn produces a force on the second charged object. The idea of an electric field might appear to be quite an artificial construction, but under some circumstances the electric field continues to exist after the charged object producing it has been removed. One therefore might conclude that the idea of the electric field has some validity beyond that of helping us visualize the electric force. In fact, this approach (associating a field with a force) has been extremely useful over the last one hundred years, successfully describing not only the electrical force, but all of the fundamental forces of nature. It should be noted that the electric field from a single charged object extends to infinity, since the electric force weakens with distance but does not become zero. All of us thus live in the electric field produced by charged objects throughout the universe.

The electric field at any point in space has a direction associated with it. By convention, it is the direction of the force experienced by a positively charged ''test'' object placed there. If lines are drawn parallel to the direction of the electric field, one obtains a schematic drawing that depicts the electric field. One such diagram is shown in figure 3.2. The direction of the electric field is given by the light lines with arrowheads on them. The direction of the force experienced by a positive charge at point *A* and a negative charge at point *B* are given by the two bold arrows. Thus electric fields can be produced in a region of space by bringing charged objects nearby.

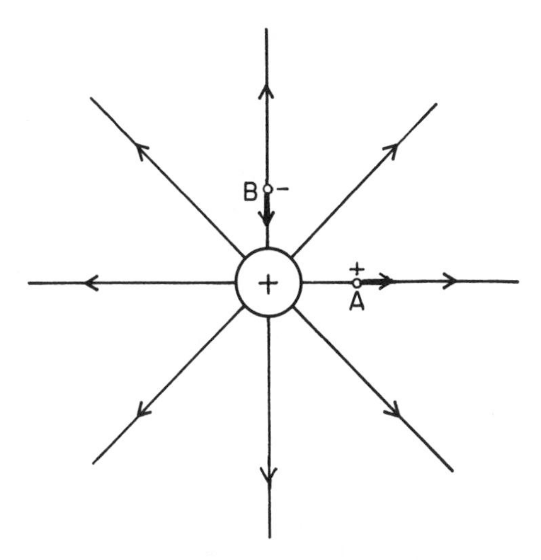

Fig. 3.2 Electric field around a positively charged object. The field lines extend outward while the force on nearby charged objects lies along the field lines (outward for positively and inward for negatively charged nearby objects).

Molecules in Electric Fields

When electric fields are applied to molecules, different kinds of molecules experience different forces. If the molecule is charged, it experiences a force in the direction of the field if it is positively charged and in the direction opposite to the field if it is negatively charged. This force tends to move the molecule along one of these two directions.

Most liquid crystal molecules are composed of neutral atoms and therefore are not charged. However, sometimes the bonding between the atoms of the molecule causes one part of the molecule to be slightly positive and another part to be equally slightly negative. This slight separation of positive and negative charge is called a *permanent electric dipole*. If no electric field is present, the permanent electric dipoles on the liquid crystal molecules are not aligned, even if orientational order is present. This is true for molecules with a slight separation of charge along the molecule because a molecule can tend to orient along the director with either the positively charged end or the negatively charged end pointing in the same direction. In the absence of an electric field, there are equal numbers of molecules with positively and negatively charged ends pointing along any direction. Likewise, a permanent electric dipole across the molecule is equally likely to orient in any direction perpendicular to the director.

The situation is very different, however, if an electric field is applied. As shown in figure 3.3, the charged parts of the molecule experience opposite forces. The two forces on the molecule do not cause it to move in one direction; rather, they tend to cause the molecule to rotate until the positive and negative parts line up with the electric field. Notice that this tends to cause the molecule to orient parallel or perpendicular to the electric field, depending on how the separation of charge occurs in the molecule. As the strength of the electric field is increased, the forces tending to orient the molecule along a certain direction increase.

In many liquid crystal molecules, the bonding of the atoms does not produce any separation of charge. These molecules respond to electric fields in a third way. Since the electric field produces forces on all of the atomic charges (positive nuclei and negative electrons), it is often possible for the electric field to displace the positive charges slightly in one direction and the negative charges in the other direction. This produces an electric dipole, but one that is only present in an electric field. Such dipoles are called *induced electric dipoles* and usually are much weaker than permanent electric dipoles.

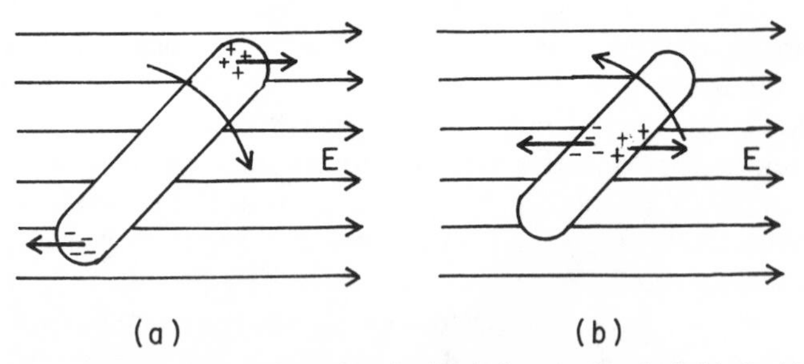

Fig. 3.3 Orientation of an electric dipole by an electric field. In (a) the dipole is along the long axis of the molecule while in (b) it lies across the long axis. The presence of the electric field causes rotation of the molecule as shown by the curved arrows.

However, an induced electric dipole experiences the same forces in an electric field as a permanent dipole and thus tends to orient itself with the positive and negative parts of the molecule along the field.

In general, liquid crystal molecules can possess permanent or induced electric dipoles both along and across the long axis of the molecule. The molecule will orient so the larger of the two electric dipoles lies along the electric field.

LIQUID CRYSTALS IN ELECTRIC FIELDS

Knowing how molecules respond to electric fields, it is not difficult to understand how liquid crystals behave. If the molecule is one that tends to orient its long axis along the electric field, then an electric field causes the molecules of the liquid crystal to tend to lie along the field. The orientational order is no greater than in the absence of an electric field; the difference is that the electric field causes the director of the liquid crystal to orient along the field. If the molecule tends to orient perpendicular to the electric field, then the presence of an electric field causes the director to lie perpendicular to the field. The strength of the electric field necessary to orient the director of a liquid crystal is relatively low since the director of a liquid crystal is usually free to orient in any direction. It is interesting to contrast this behavior to what happens in solids and liquids. It certainly is true that molecules with permanent or induced electric dipoles tend to orient along the electric field direction regardless of the phase they are in. In liquids, however, the disordered motion of the molecules overcomes any chance for them to orient. In a sense there is nothing in the phase that can orient. In solids the molecules are prevented from orienting with the field since the bonds between molecules prevent any change of orientation. The freedom of liquid crystal molecules to change orientation (like liquids) but to do so while maintaining some orientational order among the molecules (like solids) produces this delicate response to electric fields.

There is another way to view the response of liquid crystals to electric fields. Consider a sample of liquid crystal with its director locked in place relative to its container (later we will discuss ways in which this can be done). If an electric field is applied along the director, electric dipoles along the long axis of the molecules add together to make the liquid crystal sample itself be one large electric dipole. This fact is easily visualized. Consider a collection of rodlike molecules all pointing more or less up and down, each with the positively charged end up and the negatively charged end down. In the interior of the sample, the effect of the positively charged ends tend to be canceled by the negatively charged ends. At the top of the sample, however, there are more positively charged ends than negatively charged ends, so the top of the sample is positively charged. The same thing occurs in reverse at the bottom of the sample, where an overall negative charge occurs. The positive charge at the top of the sample and the negative charge at the bottom of the sample cause the liquid crystal sample itself to be one large electric dipole. Because the strength of the electric dipole depends on the size of the sample, scientists usually refer to the electric dipole per unit volume, since this quantity is independent of sample size. The electric dipole per unit volume is called the *electric polarization.*

If an electric field is applied perpendicular to the director (remember the director is fixed relative to its container), electric dipoles perpendicular to the long axis of the molecules add together to give an electric polarization along the field and perpendicular to the director. For each liquid crystal, the electric polarization along the director will be different from the electric polarization perpendicular to the director, even if the electric field used to produce them is the same (another example of anisotropy). Which electric polarization is larger depends on which electric dipole (along or across the long axis of the molecule) is larger. As we have seen, if the director is not constrained by other forces, the larger electric polarization (corresponding to the larger molecular electric dipole) tends to orient more, causing the director to point in a certain direction. The larger the difference in the two electric polarizations (in other words, the larger the anisotropy), the smaller the electric field necessary to orient the liquid crystal.

The larger the electric field applied to the liquid crystal, the larger the polarization. In many cases, the ratio of the polarization to the strength of the electric field is constant, but different depending on whether the electric field is applied parallel or perpendicular to the director (which is fixed relative to the container). This ratio is called the *electric susceptibility* and is a measure of how easily a material is polarized by an electric field. A graph of the electric susceptibility for a typical liquid crystal is shown in figure 3.4. Notice that the difference between the two electric susceptibilities decreases with increasing temperature. This is due to the decrease in the amount of orientational order as the temperature increases (see figure 1.5).

Magnetic Fields

Understanding the forces between charges becomes more complicated when the charges are moving. Not only is there an electric force between two moving charges, the motion of the charges produces an additional force called the magnetic force. Just as in the case of the electric force, two moving charges experience the magnetic force no matter what the distance between them is, with the strength of the force decreasing as the distance increases. To describe how this force works over large distances, scientists again use the concept of a field. A moving charge produces a *magnetic field* around it; when another moving charge is brought into this magnetic field, it experiences a magnetic force. Again, this concept of a magnetic field does seem to

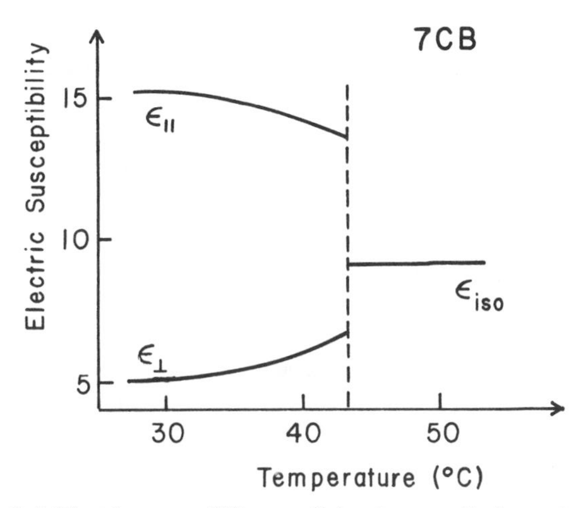

Fig. 3.4 Electric susceptibility parallel and perpendicular to the director for the liquid crystal 4-*n*-heptylcyanobiphenyl (7CB). The dashed vertical line indicates the nematic liquid crystal to isotropic liquid phase transition.

describe nature quite closely, as magnetic fields can continue to exist "on their own" after the moving charges that produced them no longer exist.

Electricity and magnetism are intimately linked, as you might guess from the fact that charges produce both effects. The two phenomena are very similar. Just as there are two types of charges that produce opposite electric effects, opposite magnetic effects are produced by a positive charge moving in one direction and a positive charge moving in the opposite direction. Furthermore, a positive charge going in one direction and a negative charge going in the same direction also produce opposite magnetic effects. Thus there are only two types of moving charges as far as magnetism is concerned, just as is true for electricity.

One important difference between electricity and magnetism can now be appreciated. To produce magnetic fields that exist for some period of time, charges must move for a similar period of time. To keep charges moving requires that they travel around a closed loop of some kind, constantly receiving energy from a device in the "circuit" in order to keep moving. It is therefore impossible to construct what would be the theoretically simplest magnetic situation, that of charge moving in only one direction. Instead, one must consider the magnetic effects produced by a small loop of moving charge as the starting point. A diagram of the magnetic field for such a loop is shown in figure 3.5. Notice that the lines which represent the magnetic field are quite different from the electric field lines of figure 3.2. To help describe the magnetic field for a loop, scientists label the region to the right of the loop (where the field lines diverge from the loop) the north pole of the loop, and the region to the left of the loop (where the field lines converge into the loop) the south pole. Ordinary bar magnets act like a series of loops placed next to each other; the magnetic field lines emerge from one end of the set of loops (the north pole) and converge into the other end of the set of loops (the south pole).

The fact that the basic "unit" of magnetism is charge flowing around a loop makes the basic interaction of magnetism different from electricity. A loop of moving charge placed in a magnetic field does not experience a net force, because the loop has charge flowing in many directions. There are forces on different parts of the loop,

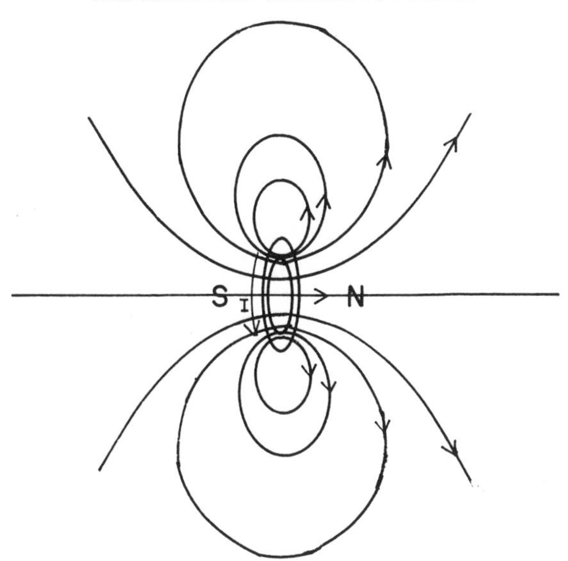

Fig. 3.5 Magnetic field around a loop of moving charge. The direction of positive charge movement is given by the arrow (labeled *I*). The north and south poles are indicated by the letters *N* and *S*, respectively.

but they add up to zero for the loop as a whole (see figure 3.6[a]). However, if the loop is not facing "broadside" to the magnetic field, these forces on different parts of the loop tend to orient the loop so it is "broadside." This effect is shown in figure 3.6(b). This situation is analogous to what happens to an electric dipole in an electric field. In fact, a small loop of moving charge is sometimes called a *magnetic dipole*.

Molecules and Liquid Crystals in Magnetic Fields

Although we have been using large loops of moving charge as our example, it is important to understand that nature provides us with loops of moving charge on the atomic scale. The negatively charged electrons around the nucleus of an atom sometimes behave as tiny loops of moving charge. Some atoms act as *permanent magnetic dipoles*, in effect never-ending loops of moving charge. These atoms

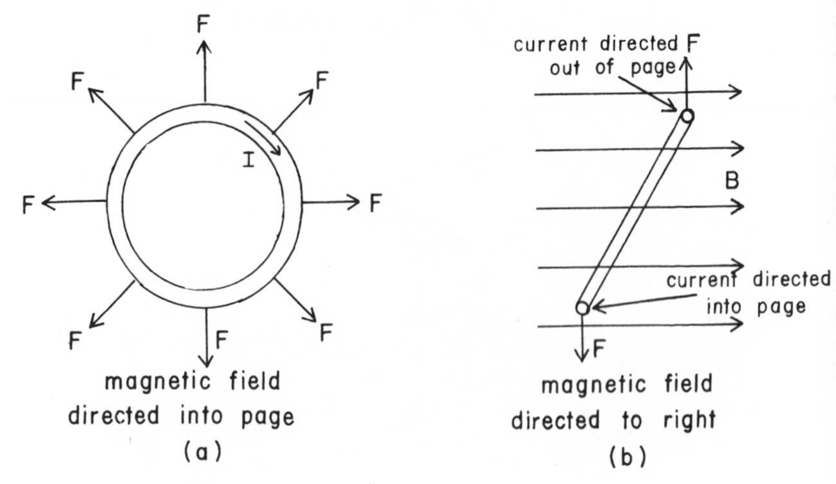

Fig. 3.6 Forces on a loop of moving charge in a magnetic field. The direction of positive charge movement is given by the arrow labeled *I*. With the loop broadside to the field (a), the forces on the loop cancel each other and produce no net force. With the loop not broadside to the field (b), the forces do not cancel and cause the loop to orient broadside to the field.

are the ones that can be used to make permanent magnets. The atoms ordinarily in liquid crystal molecules are not of this type, but just as in the electric case, a magnetic field applied to the molecule will cause some of the charges within the molecule to act as tiny loops of moving charge. Once this happens, the molecule possesses an *induced magnetic dipole*, and this dipole will tend to orient with its north and south pole along the magnetic field. The atoms present and how they are bound together in a molecule cause the induced magnetic dipoles to differ from one molecule to the next. Induced magnetic dipoles both parallel and perpendicular to the long axis of the molecule are possible. Liquid crystal molecules therefore tend to align either parallel or perpendicular to the magnetic field.

Just as in the electric case, the presence of a magnetic field causes a liquid crystal sample itself to possess a magnetic dipole. The magnetic dipole per unit volume of liquid crystal is called its *magnetization*. In a fashion similar to the electric case, a magnetic field applied parallel to the director (assumed locked in position relative to its container) induces a certain amount of magnetization, while a magnetic field applied perpendicular to the director induces a different amount of magnetization. This is simply another example of anisotropy in liquid crystals, which originates in the different magnetic dipoles induced along or across the long axis of the molecule. If the director is not constrained by other forces, the liquid crystal will tend to orient so the higher magnetization (corresponding to the larger molecular magnetic dipole) is parallel to the field. Again, since the director of a liquid crystal is normally not locked in place relative to its container, the director can easily change its orientation. Liquid crystals are therefore very sensitive to magnetic fields, achieving complete alignment of the director for magnetic fields of relatively low strength. The higher the anisotropy of the magnetization in the liquid crystal, the smaller the magnetic field necessary to align the director.

Just as in the case of an electric field, a higher magnetic field strength produces a higher magnetization. The ratio of the magnetization to the strength of the applied magnetic field is usually constant and is called the *magnetic susceptibility*. Liquid crystals have two magnetic susceptibilities: one for magnetic fields applied parallel to the director and one for magnetic fields applied perpendicular to the director (assuming the director is fixed relative to the container). These vary with temperature in liquid crystals in a way similar to the electric susceptibilities, since both depend on how the order parameter varies with temperature. Data for a typical liquid crystal are shown in figure 3.7.

Deformation of Liquid Crystals

Although the director in a liquid crystal frequently is free to point in any direction, there are some important examples where this is not the case. Early scientists were quick to notice that some materials, when placed in contact with a liquid crystal, force the director to point along a specific direction. For example, a glass surface can be prepared in such a way that the liquid crystal molecules near the glass are forced to lie parallel to the surface. The director must therefore lie parallel to the surface, and will remain in this configuration even if an electric or magnetic field is applied. The molecules farther away from the surface are not constrained in this way, so they can respond

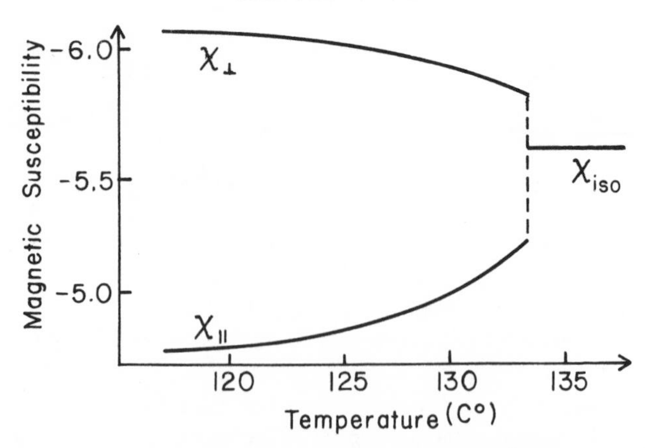

Fig. 3.7 Magnetic susceptibility parallel and perpendicular to the director for the liquid crystal *p*-azoxyanisole (PAA). The dashed vertical line indicates the transition from nematic liquid crystal to isotropic liquid.

to electric and magnetic fields. The result of applying a field is to cause the liquid crystal to deform, with the director parallel to the surface in the vicinity of the surface and perpendicular to the surface away from the surface (assuming a field causing alignment of the director perpendicular to the surface). In this situation, the effects of the surface and electric or magnetic field compete, causing deformation of the liquid crystal. To understand how liquid crystals respond to electric and magnetic fields, therefore, we must also discuss effects related to deformation.

An undeformed liquid crystal is one in which the director points in the same direction throughout the liquid crystal. This is the director configuration that the liquid crystal adopts if it is free from other influences. A deformed liquid crystal is one in which the director changes its direction from point to point. In describing all the ways this might occur, scientists have discovered that all deformations in liquid crystals can be described in terms of three basic types of deformation: (1) *splay*, (2) *twist*, and (3) *bend*. These basic deformations are shown in figure 3.8. For all three deformations shown, the director is unchanged in moving into or out of the page. Thus these three basic deformations involve director orientation changes in only two dimensions. In cases of a complicated deformation, more than

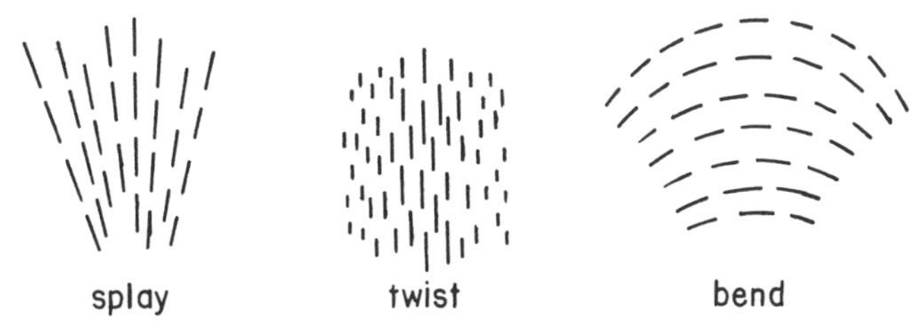

Fig. 3.8 The three basic types of deformation in liquid crystals. In all three cases, deformation is present in the plane of the page. There is no chage in the director orientation in going into or out of the page.

one of these basic deformations must be used to describe it (for example, both splay and bend in the same plane).

In trying to understand how liquid crystals deform under various conditions, it is useful to think of a deformed liquid crystal as a compressed spring. An outside force is necessary to deform a liquid crystal just as an outside force is necessary to compress a spring. Once this force is removed, the liquid crystal returns to its undeformed configuration just as the spring returns to its normal length. The force required to deform a liquid crystal increases as the amount of deformation increases much like the force required to compress a spring grows as the spring is compressed more and more. When an electric or magnetic field deforms a liquid crystal, the field must supply the force necessary to cause the deformation.

One exception should be pointed out at this time. The molecules that form chiral nematic liquid crystals possess forces between them that on average cause the molecules to orient at a slight angle with respect to each other. This ordering of the molecules causes the director to rotate, giving the liquid crystal a certain amount of twist. For chiral nematic liquid crystals, therefore, the twisted configuration is the spontaneous one; a force is required to produce splay or bend or to change the amount of twist from its spontaneous value.

Effects in Thin Samples of Liquid Crystals

To study liquid crystals, scientists often place a sample between two pieces of glass. This serves to keep the liquid crystal contained in one place, while allowing plenty of light to enter and leave the sample. Although very convenient, this procedure causes the liquid crystal to behave in a complex way, due to the competition between many effects. In this situation, consideration must be given to both the electric or magnetic field effects and the deformation properties in order to gain an understanding of how the liquid crystal behaves.

So delicate is the response of liquid crystal molecules that almost any surface causes the director to orient in a specific direction near the surface. For example, a glass surface need only be rubbed with a piece of cloth before liquid crystal is placed against it in order to force the director to lie along the direction of rubbing. Other effective treatments include the application of a chemical to the glass before adding the liquid crystal. The chemical (called a surfactant) adheres to the glass and produces alignment of the director. Evaporation of solid material onto the glass prior to using it with a liquid crystal also causes alignment of the director. In short, proper preparation of the glass surface can produce alignment of the director in almost any direction. The discussion here is limited to two cases, alignment parallel and perpendicular to the surface of the glass.

Consider what happens when a small amount of nematic liquid crystal is placed between two pieces of glass that have been treated to produce alignment of the director parallel to the surface. Near the two glass surfaces, the director is constrained to point in a certain direction (parallel to the surface). To remain in an undeformed configuration, the liquid crystal between the glass surfaces also adopts a configuration where the director is parallel to the surfaces. This configuration is shown in figure 3.9(a) and is called the *homogeneous texture*. Now imagine that an electric or magnetic field is applied perpendicular to the glass surfaces and that this field tends to orient the director parallel to the field. The molecules near the surfaces are not very free to orient with the field, but the molecules near the middle are fairly free. The electric or magnetic field thus causes the director to change its orientation the most in the middle, with diminishing change closer to the surfaces. This deformed structure is shown in figure 3.9(b). The most interesting aspect of this deformation is that it does not occur gradually as the strength of the electric or magnetic field is gradually increased. For fields with strengths below a certain value, the liquid crystal remains in its completely undeformed homogeneous texture. No change at all occurs. Then at some thresh-

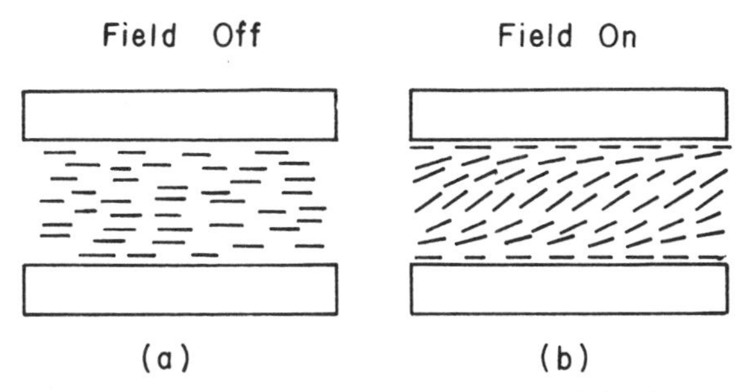

Fig. 3.9 Freedericksz transition in the splay and bend geometry. Treatment of the glass orients the liquid crystal parallel to the surfaces, while the field tends to align the director perpendicular to the surfaces. The condition when the field is below the threshold value is shown in (a); (b) shows the condition when the field is above the threshold value.

old value of the field, the deformation begins and then gets greater as the field strength is increased. This transition from an undeformed to deformed texture at a certain value of the field is called the *Freedericksz transition*, after the Russian scientist who studied this phenomenon in the 1930s. The Freedericksz transition is not a phase transition, because at any point in the liquid crystal the order of the molecules relative to one another remains the same. It is simply a transition from a uniform director configuration to a deformed director configuration.

From the previous discussion, it is easy to guess what factors must enter into determining the value of the threshold field. Since the anisotropy of the electric polarization or magnetization determines how much tendency there is for the director to orient along the field, the higher the anisotropy of the liquid crystal, the lower the value of the threshold field. Likewise, liquid crystals differ in the magnitude of the force necessary to deform them. Liquid crystals requiring higher forces for deformation obviously have higher threshold field values. Finally, since the deformation occurs continuously across the thickness of the liquid crystal film, the greater the thickness, the less the deformation (change in director orientation per unit distance) at any point. Since the same alignment of the director in the center of the film means less deformation in thicker films, one expects the threshold value for the field to be less in thicker samples. This turns out to be the case. Notice that the first two parameters are properties of the liquid crystal. Once a liquid crystal is chosen for the thin film, these two parameters are set. Only the thickness can then be varied.

The deformation shown in figure 3.9(b) contains both splay and bend. By treating the surfaces of the glass differently and by directing the electric or magnetic field in other directions, it is possible to produce other types of deformations. Consider as a second geometry the case shown in figure 3.10. Again the surfaces have been treated for parallel alignment of the director, but now the field is applied into or out of the page. The molecules in the middle of the liquid crystal film tend to point in or out of the page with the field, so the deformed state is one of twist. This is shown in figure 3.10(b). Although the geometry of this situation is different, the changes occur much like in the previous geometry. No deformation takes place at all until the electric or magnetic field strength reaches a certain threshold value, at which time the deformation begins. The value of the threshold field strength depends on the same factors as before, except now the force necessary to cause twist deformation is important, rather than splay and bend deformation.

The third different geometry results from treating the glass surfaces in a way that produces perpendicular alignment of the director. Again, with no electric or magnetic field applied, the liquid crystal adopts a completely undeformed director configuration that always

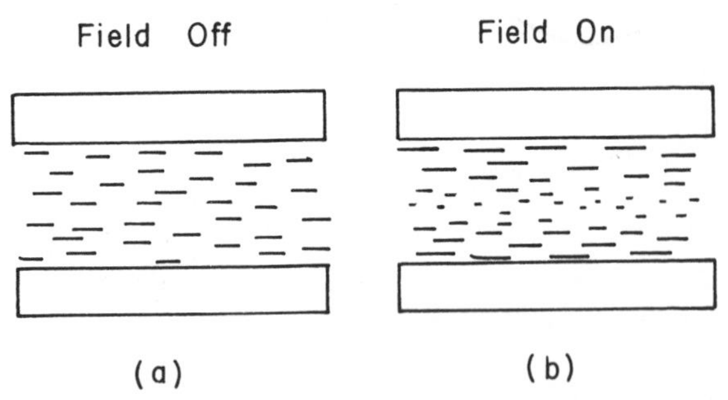

Fig. 3.10 Freedericksz transition in the twist geometry. Treatment of the glass orients the liquid crystal parallel to the surfaces and in the plane of the page, while the field tends to align the director perpendicular to the page. The condition with the field below and above the threshold value is shown in (a) and (b), respectively.

points perpendicular to the surfaces. This is shown in figure 3.11(a) and is called the *homeotropic texture*. The electric and magnetic field is applied parallel to the glass surfaces, thus tending to orient the molecules in the middle parallel to the surfaces. The deformed state is shown in figure 3.11(b); notice that the deformation is purely bend. As in the other two geometries, no deformation results as the field is increased until it reaches a threshold value. The value of this threshold field depends on the anisotropy and film thickness as before, but is dependent only on the force necessary to produce bend deformation in the liquid crystal. The transition from undeformed to deformed director configuration is called a Freedericksz transition in all three cases.

One important aspect of the Freedericksz transition is that typical values for the threshold field are not very high. This is due to the relative freedom liquid crystal molecules possess as they diffuse throughout the sample. For a 25 micrometer (0.001 inch) thick film, the threshold electric field is roughly 400 volts/centimeter and the threshold magnetic field is about 0.2 teslas. Both of these field strengths are easily produced in the laboratory. In fact, the threshold electric field can be produced by applying only 1 volt to electrodes attached to the glass surfaces. The Freedericksz transition is important to the operation of liquid crystal displays, because these transitions cause dramatic changes in the optical characteristics of the liq-

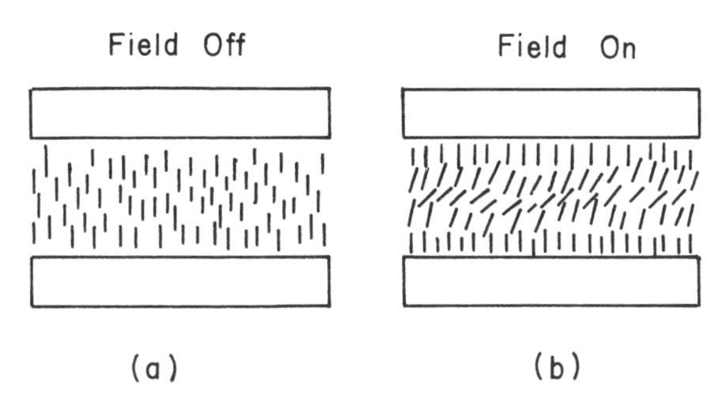

Fig. 3.11 Freedericksz transition in the bend geometry. Treatment of the glass orients the liquid crystal perpendicular to the surfaces, while the field tends to align the director parallel to the surfaces. The conditions below (a) and above (b) the threshold field strength are both shown.

uid crystal film. How light interacts with liquid crystals is one of the most interesting subjects in the field of liquid crystals, and is discussed at length in the next chapter.

Hydrodynamic Effects Caused by Electric Fields

One last example of how liquid crystals delicately respond to electric or magnetic fields is the flow patterns that develop under certain conditions. Motion of the fluid induced by electric fields was noticed by some of the early researchers, but was not well understood until the 1960s. The reason for this is that liquid crystal molecules are uncharged, and therefore should not experience a force in an electric field. An electric field may orient the liquid crystal molecules, but not cause motion of the entire molecule in a certain direction. What then causes the motion?

The answer lies in the presence of a small amount of charged impurity ions in the liquid crystal sample. The electric field causes these ions to move from one electrode to the other, but the mobility of these ions is greater parallel to the long axis of the molecules than perpendicular to it. Under some conditions this movement of ions can actually orient the liquid crystal molecules along the electric field, even if an orientation perpendicular to the electric field would otherwise be preferred. In other cases a regular pattern of bright and dark stripes (called *Williams domains*) is produced and dust particles being carried along by the moving liquid crystal molecules can be seen undergoing periodic motion within the striped bands. This is a more complicated situation, since the interaction between the motion of the ions and the liquid crystal produces a slight separation of the positive and negative moving ions. As shown in figure 3.12, convection cells (shown by the arrows) result as the molecules flow in response to this slight separation of the differently charged ions (shown as circles). The convection cells cause a regular distortion of the director throughout the sample, which in turn serves to focus light slightly at regularly spaced points in the liquid crystal. It is this slight focusing of light that produces the pattern of stripes.

If the electric field is strong enough, the flow becomes turbulent. No longer is the flow steady and regular; instead the motion of the

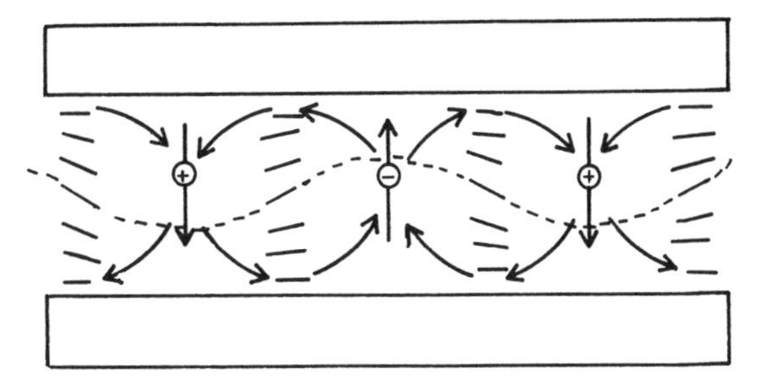

Fig. 3.12 Formation of Williams domains in a liquid crystal cell. Motion of charged impurities (circles) causes the liquid crystal molecules to circulate (arrows). This circulation causes a distortion of the director configuration (shown by the dashed line), which focuses light through the sample into alternating bright and dark bands.

molecules at any point in the sample changes erratically. This much more violent motion causes intense scattering of light, and thus the sample appears much brighter and whiter. Scientists refer to this phenomenon as *dynamic scattering*, and it has been the basis for a number of useful applications.

CHAPTER 4

Light and Nematic Liquid Crystals

The optical properties of liquid crystals are one of the most interesting and certainly the most beautiful features of the phase. How liquid crystals effect light is also the basis for just about all of the applications of liquid crystals. In this chapter we will come to understand what happens to light as it propagates through a sample of nematic liquid crystal. To do this properly, we must first discuss some of the properties of light itself.

LIGHT

Although light has always been a topic of much scrutiny for the civilizations of the world, an understanding of what light actually is only became possible in the late nineteenth century after the development of a new theory for electromagnetism. Our discussion about electric and magnetic fields and how matter behaves in the presence of these fields was based on these same theoretical ideas. Perhaps the most interesting aspect of this theory is the prediction that oscillating electric and magnetic fields can propagate through space without the need for some kind of medium. That is, these oscillating electric and magnetic fields can travel through a perfect vacuum. The scientists of the late nineteenth century saw the connection between this prediction and all the properties of light, so the idea of light as propagating electric and magnetic fields (that is, an *electromagnetic wave*) quickly became a widely accepted notion.

The theory predicts that the electric and magnetic fields in an electromagnetic wave must have certain characteristics. The simplest example is a wave in which the electric and magnetic fields are perpendicular to each other, with both fields perpendicular to the direction the wave is propagating. The strength of both the electric and magnetic fields oscillates as the wave propagates, reaching a maximum value in one direction, then decreasing through zero until it reaches a maximum value in the opposite direction, then decreasing through zero again, etc. Such an electromagnetic wave is shown in figure 4.1, where a set of axes has also been drawn for reference. This wave has the electric field always directed along the x-axis and the magnetic field always directed along the y-axis, while the wave propagates along the positive z-axis.

There are two important characteristics of electromagnetic waves that are evident from the figure. First of all, notice that the electric and magnetic fields cycle through their values quite regularly. This means that any value of the electric or magnetic field repeats itself after a certain distance along the z-axis. This is shown in figure 4.1 for two places where the electric field reaches a maximum along the positive x-axis. This distance is an important characteristic of any wave and is called its *wavelength*. The distance between two successive crests in an ocean wave is a measure of its wavelength in an analogous fashion. The second important characteristic stems from the fact that the electric field of the wave in figure 4.1 points along a single direction in space. Since the magnetic field must be perpendicular to the electric field, this implies that the magnetic field also points along a single direction in space. This is a special case, so we say that the wave is *linearly polarized*. In stating the direction of *light polarization*, we adopt the convention of choosing the direction of

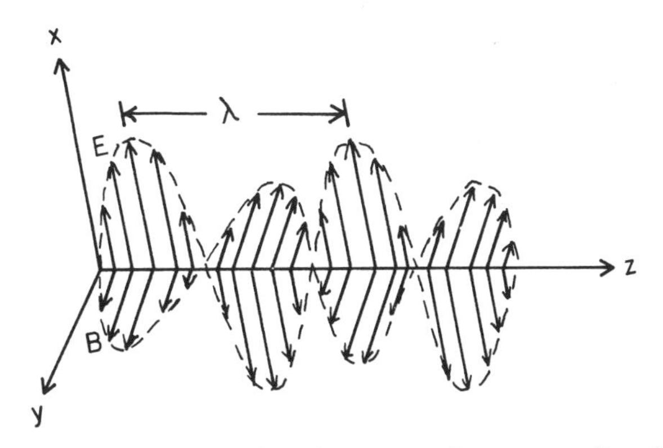

Fig. 4.1 Representation of an electromagnetic wave traveling along the z-axis. The electric field is always parallel to the x-axis and the magnetic field is always parallel to the y-axis. The wavelength is denoted by the symbol λ.

the electric field; the wave in figure 4.1 is linearly polarized along the x-axis.

A word of caution is necessary. In the last chapter we described the "polarization of a sample" as the electric dipole per unit volume. This "polarization" resulted from the application of an electric field to the sample. Here we talk about the "polarization of light" as describing the nature of the electric and magnetic fields of the light. Although scientists use the same word to describe these two phenomena, the two ideas are completely separate. In all cases, the proper meaning of the word "polarization" will be evident from the situation under discussion.

The wavelength of an electromagnetic wave can be as long as hundreds of miles or shorter than the dimensions of a proton. Electromagnetic waves of different wavelengths behave in very different ways, so special names have been given to electromagnetic waves depending on the wavelength. For example, electromagnetic waves with a wavelength greater than 1 meter are called radio waves. Microwaves have wavelengths between 1 millimeter and 1 meter, while X-rays have wavelengths between 0.01 and 1 nanometer (1 nanometer equals 0.000000001 meter). What differentiates light from all other electromagnetic waves is that our eyes are only sensitive to electromagnetic waves with a small range of wavelengths, between 400 and 700 nanometers. Electromagnetic waves with a wavelength in this range constitute light waves and our eyes can detect them. Within this range of wavelength, our eyes respond differently to each wavelength, thus causing the sensation of color. An electromagnetic wave with a wavelength of 420 nanometers appears violet to us, while one of 470 nanometers appears blue, 530 nanometers appears green, 580 nanometers appears yellow, 610 nanometers appears orange, and 660 nanometers appears red. We cannot see electromagnetic waves with a wavelength slightly shorter than 400 nanometers or slightly longer than 700 nanometers, so we call them *ultraviolet* and *infrared* waves, respectively.

Polarized Light

The linearly polarized wave shown in figure 4.1 is only one example of how electromagnetic waves (light) can be polarized. Another equally simple example would be a wave linearly polarized along the y-axis. In this case the electric field always points along the y-axis and the magnetic field always points along the x-axis. More complicated examples of polarized light can be understood as a combination of light linearly polarized along the x-axis and light linearly polarized along the y-axis. Before we can appreciate this, however, we must first consider what happens when two electric fields are present simultaneously.

Recall that the concept of the electric field was really a way of describing the electric force. We can therefore create a situation with an electric field along the x-axis and an electric field along the y-axis by situating two charges in the proper locations and asking about the force on a third charge (called a "test" charge). This is shown in figure 4.2(a), where the two arrows represent the two electric forces on the "test" charge due to the two large charges. The arrows are of different lengths because the forces due to each of the larger charges are not equal. The total force on the smaller charge is just the sum of both forces, making sure we take the direction of the two forces into account. A convenient way of doing this is to construct the rectangle defined by the two forces, since the total force is just along the diagonal. This procedure is shown in figure 4.2(b). If we wish to describe the electric field at the location of the smaller charge, therefore, we

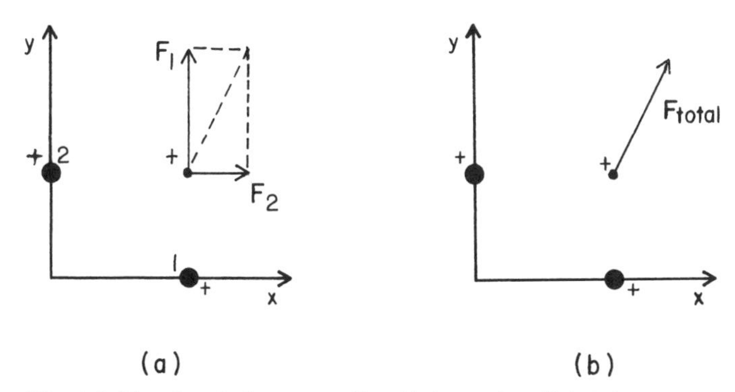

Fig. 4.2 The electric force on a "test" charge (small dot) due to two other charges (large dots). F_1 is the force due to charge 1 and F_2 is the force due to charge 2. The total force (sum of F_1 and F_2) is shown in (b).

would simply say it points along the direction of the total force in figure 4.2(b), since by definition the direction of the electric field is in the same direction as the electric force.

We can also arrive at the same result through a consideration of the electric fields due to the two charges rather than the force they produce. Figure 4.3(a) shows the electric fields due to the two larger charges at the point in space where the "test" charge is located. The different lengths of the two arrows again indicate that the strengths of the two fields are not equal. Notice that forming the rectangle using these two fields (see figure 4.3[b]) produces the same electric field direction as before. Therefore, when two electric fields are present along the x-axis and y-axis, the total electric field can be obtained by drawing the diagonal for the rectangle formed by the two electric fields.

The procedure just described is valid for adding together any two quantities that possess directions at right angles to one another. For example, imagine you are in a large city and go for a walk. First you walk four blocks east. On a map of the city, this walk can be represented by an arrow to the right four blocks long. Next you walk five blocks north. This is represented by an arrow five blocks long pointing toward the top of the map. The addition of these two "walks" can be represented by an arrow from your starting position on the map to your ending position. This is simply the diagonal of the rectangle formed by the east and north pointing arrows. You can prove

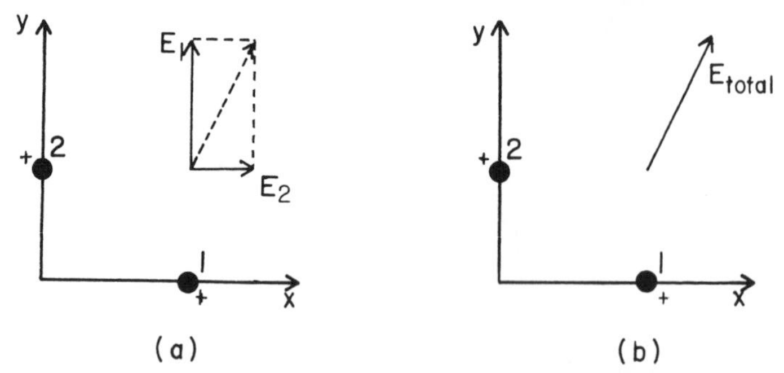

Fig. 4.3 The electric field at a point in space due to two charges. E_1 is the electric field due to charge 1 and E_2 is the electric field due to charge 2. The total electric field (sum of E_1 and E_2) is shown in (b).

to yourself that the ''walk'' represented by this diagonal actually equals the sum of the two other ''walks,'' because you know that if you could take this diagonal ''walk'' (the buildings do present somewhat of a hazard), you will end up at the same position as after taking the east and north ''walks.'' In analogous fashion, the diagonal electric field is identical to the sum of the two fields along the x-axis and y-axis.

This technique can be used to determine what happens when linearly polarized light along both the x and y directions is present. Let us first examine the case when the maximum strengths of the two electric fields are equal. That is, the electric fields of both waves oscillate, but the maximum value of the electric field along the x-axis and the maximum value of the electric field along the y-axis are equal. Now imagine sitting at a point along the z-axis as these two electromagnetic waves pass. The electric field of both waves will oscillate as the waves pass, starting from zero, reaching a maximum in one direction, going through zero again, reaching a maximum in the other direction, and so on. To understand this situation, we must keep track of what the electric field along both the x-axis and y-axis is doing. Figure 4.4(a) shows the strength of both electric fields at eight different times as the two waves pass. The numbers on each of these individual electric fields give the chronological order. That is, the electric fields of both waves start at zero strength (1), reach a maximum in one direction (3), pass through zero again (5), reach a maximum in the other direction (7), and finally reach zero again. The electric fields at times intermediate to these times are also shown. Let us add together the two electric fields at each of these eight points in time. This is shown in figure 4.4(b), where the different rectangles have been drawn. The numbers again give the chronological order. Notice that the electric field simply oscillates along a line at 45° to the x-axis and y-axis. The combination of these two linear polarizations is just linearly polarized light along a different direction.

In the previous example, the electric fields of the linearly polarized waves both started out at zero and increased in the positive x and y directions. Consider what occurs if they start out at zero again, but with the light linearly polarized along the y-axis now increasing in the opposite direction. Figure 4.5(a) shows the individual electric fields at the same eight times and figure 4.5(b) gives the result. The

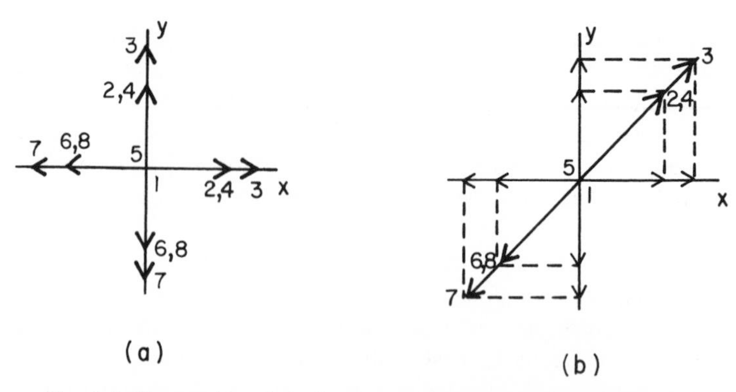

Fig. 4.4 The combination of two electromagnetic waves with polarizations along the x-axis and y-axis. The electric field strengths for each of the waves are given in (a). The numbers represent different times in chronological order. The combined electric field strengths are given in (b) for the same eight times. The phase difference between the two waves is 0°.

numbers again denote the chronological order as the two waves pass. The result is again linearly polarized light, but now polarized along a different direction.

Consider one more case. What happens if the electric field polarized along the x-axis reaches its maximum value along the positive x-axis just as the electric field polarized along the y-axis starts to increase from zero along the positive y-axis? We can apply the same procedure as before, but the result is very different. As can be seen from figure 4.6, the total electric field does not oscillate along a single direction, but rotates counterclockwise, always maintaining the same strength. We call this *circularly polarized* light, as opposed to linearly polarized light. To indicate that circularly polarized light can rotate in either of two directions, we call light with an electric field rotating counterclockwise at some point in space (as seen by an observer facing the source of light) left circularly polarized light, and light with an electric field rotating clockwise (as seen by a similar observer) right circularly polarized light. To produce right circularly polarized light, the electric field polarized along the y-axis must reach its maximum value along the positive y-axis just as the electric field

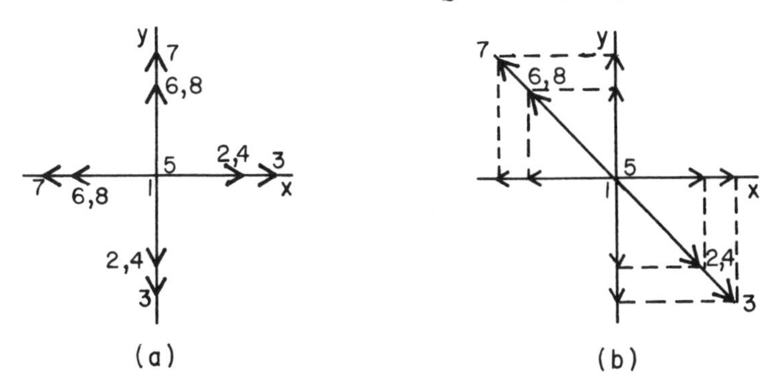

Fig. 4.5 The combination of two electromagnetic waves with polarizations along the x-axis and y-axis. The electric field strengths for each of the waves are given in (a). The numbers represent different times in chronological order. The combined electric field strengths are given in (b) for the same eight times. The phase difference between the two waves is 180°.

polarized along the x-axis starts to increase from zero along the positive x-axis.

These last four examples differ only in what each of the two linearly polarized waves are doing relative to each other. We use the concept of *phase difference* to describe these differences. For example, in the case shown in figure 4.4 both polarized waves are doing the same thing at the same time. We say there is zero phase difference between these two waves, or, alternatively, we say the two waves are in phase. In the example of figure 4.5, the electric field polarized along the y-axis is half a cycle ahead of the electric field polarized along the x-axis. Since a full cycle can be represented by 360°, we say there is a 180° phase difference between the two waves; or, since the two waves are behaving in opposite fashion, we can also say the two waves are out of phase. The example of left circularly polarized light (figure 4.6) by the same reasoning represents two waves 90° out of phase. Whether the two waves combine to give right or left circularly polarized light depends on which electric field is a quarter cycle ahead of the other.

In all of these examples, the maximum strengths of the two linearly polarized electric fields are equal. When this is not the case, (1) two waves in or out of phase give linearly polarized light at some angle

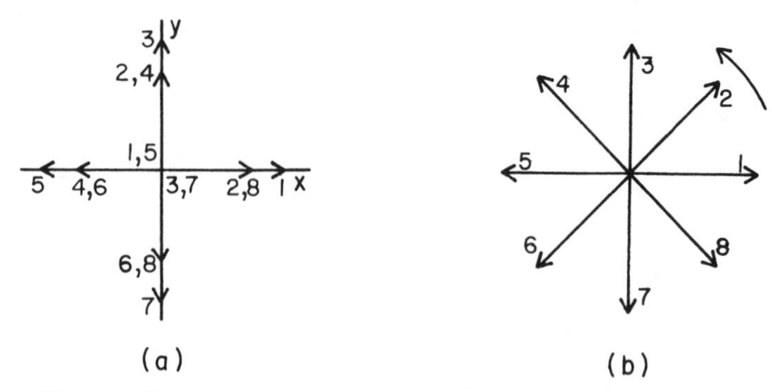

Fig. 4.6 The combination of two electromagnetic waves with polarizations along the x-axis and y-axis. The electric field strengths for each of the waves are given in (a). The numbers represent different times in chronological order. The combined electric field strengths are given in (b) for the same eight times. The phase difference between the two waves is 90°.

other than 45° to the x-axis and y-axis, and (2) two waves 90° out of phase give elliptically polarized light (the electric field strength varies as it rotates). In the latter case, the ellipse outlined by the electric field as it rotates has its long axis along the x-axis or y-axis. A phase difference of other than 0°, 90°, or 180° also produces elliptically polarized light, but now the long axis of the ellipse will not be along either the x-axis or y-axis. All three of these cases are shown in figure 4.7.

How Light Interacts with Matter

Because my discussion of light as an electromagnetic wave has assumed there is nothing around to affect the wave, the results really apply to light propagating through a vacuum. In this case, light travels at a speed of 300,000,000 meters per second or 186,000 miles per second. The presence of matter, however, affects the electromagnetic wave in several important ways. Just as a steady electric or magnetic field causes matter to become polarized or magnetized, the oscillating electric and magnetic field of a light wave produces oscillating motion of the charges in a material. These oscillating charges in turn produce their own oscillating electric and magnetic fields, resulting

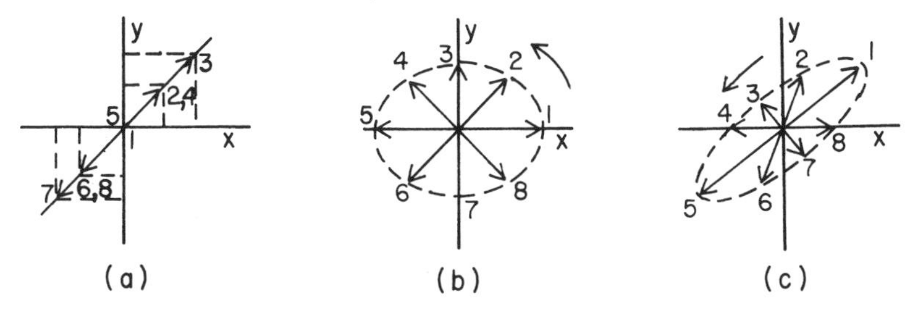

Fig. 4.7 The combination of two electromagnetic waves with polarizations along the x-axis and y-axis as in figures 4.4–4.6, but with unequal maximum electric field strengths. The phase difference between the two waves is 0° in (a), 90° in (b), and between 0° and 90° in (c).

in a combined electromagnetic wave different from the one that would have been present if no matter were around. Although this situation is in principle quite complicated, the additional electric and magnetic fields coming from all the different charges in the material cancel themselves out, except for those fields that are propagating in the same direction as the original wave. These induced electromagnetic waves combine with the original wave to produce an electromagnetic wave that travels through the material with a velocity slower than the velocity of light in a vacuum. How much slower light propagates in a material depends on the exact nature of the charges in the material. We therefore assign a number to each material that represents the factor by which the velocity of light is slowed down. That is, if v is the velocity of light in a material and c is the velocity of light in a vacuum, then this factor, n, (called the *index of refraction* of the material) is defined as

$$n = c/v.$$

It is clear from this definition that the index of refraction of all materials must be greater than one. For example, the index of refraction is 1.0003 for air, 1.33 for water, and about 1.5 for glass.

Because light of various wavelengths (as measured in a vacuum before it enters matter) produces different effects on the charges present in matter, the velocity at which light propagates through a material depends on the wavelength of the incident light. This means that the index of refraction also depends on wavelength. Usually this effect is small; for example, glass might have an index of refraction of 1.51 for red light and 1.53 for violet light. In some cases, however, this small effect can lead to dramatic results.

Light traveling through a material also tends to lose some of its intensity as the material absorbs some of the energy of the electromagnetic wave. The ability of a material to absorb some of the energy in the light wave also depends on the wavelength of the incident light, but in this case the effect can be quite large. For example, a piece of red glass in a stained glass window absorbs much more light of shorter wavelength (blue and green) than of longer wavelength (red). Although light of many wavelengths (white light) strikes the glass, much more of the red light gets through so we see a red color. The absorption properties of materials vary considerably, resulting in a wealth of possible effects.

The interaction of light with matter is even more interesting at the point where light goes from one material to another. Because the arrangement of charges in the two materials are different, the oscillating electric and magnetic fields from these charges are different. Although these fields cancel themselves except in the forward direction within each material, they do not cancel in the same way at the boundary between the two materials, because each material produces different oscillating electric and magnetic fields. The oscillating electric and magnetic fields produced at the boundary between two materials together with the original electromagnetic wave now cancel except for those traveling in two directions. One of these directions is backward at an angle, thus producing a *reflected* wave. The second direction is forward, but in a direction different from the direction of the wave in the first material. This is illustrated in figure 4.8. The bending of light as it passes from one material to another is called *refraction*. For a given angle between the boundary and the direction of propagation of the wave in the first material, the new angle in the second material depends only on the indices of refraction of the two materials. The amount of light reflected as opposed to refracted at the boundary, however, also depends on how the light is polarized.

Birefringence or Double Refraction

In the preceding discussion, we assigned a single index of refraction to each material. A simple experiment quickly indicates that this

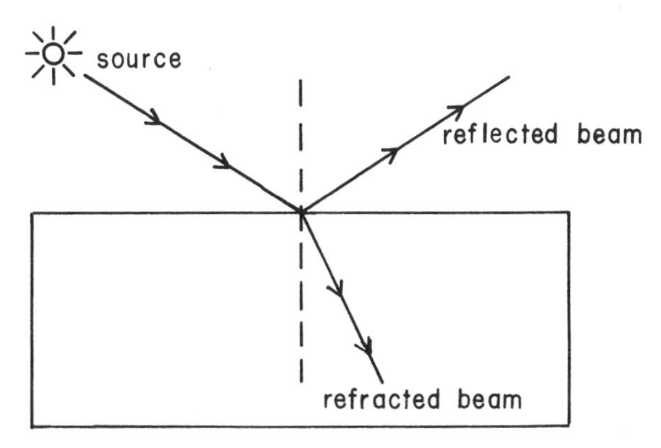

Fig. 4.8 Reflection and refraction of light at the surface of an isotropic material. The path of the reflected or refracted light is independent of the polarization of the light.

is not always valid. Let us take a beam of light that is linearly polarized along the x-axis and direct it at an object, measuring the angle the beam makes on either side of the boundary between the air and the object. Measurement of these angles allows us to find the index of refraction for the object. Just to be sure everything is working well, let us repeat the measurement for each object using light linearly polarized along the y-axis. For many objects (glass, water, a large salt crystal), both measurements give the same number for the index of refraction, but other objects (a large crystal of quartz or calcite) produce two different numbers. This means that light propagating through these latter materials is traveling at a different velocity depending on whether it is polarized along the x-axis or y-axis. As previously discussed in chapter 3, a material that has different properties for different directions is called an anisotropic material. Obviously, the first set of materials must be isotropic and possess one index of refraction, while the second set must be anisotropic and possess more than one index of refraction. This effect is easily seen when a beam of light containing both x-polarized and y-polarized light is incident on an anisotropic material. The two polarizations have different indices of refraction and therefore travel along two different directions inside the material. This phenomenon is called *birefringence* or *double refraction* and is illustrated in figure 4.9.

All effects that depend on the index of refraction are different for

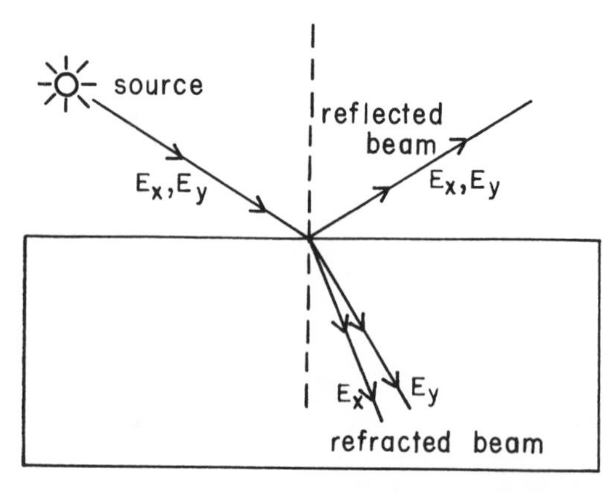

Fig. 4.9 Reflection and refraction of light at the surface of an anisotropic material. The path of the reflected light is independent of the polarization of the light. The path of the refracted light depends on whether the light is linearly polarized along the y-axis (out of the page) or x-axis (perpendicular to both the y-axis and the direction of light propagation).

these two polarizations. For example, the amount of light reflected or transmitted at the interface depends on the polarization of the light. Likewise, the amount of light absorbed by the material is different for the two polarizations. In fact, it is possible to fabricate materials that absorb one polarization nearly completely while allowing most of the other polarization to pass through. Such materials can be used to produce linearly polarized light, since only one polarization emerges when they are illuminated by unpolarized light. Similarly, these ''polarizers'' can also be used to check if a beam of light is polarized, in that one orientation of the ''analyzer'' blocks all the light if the beam is linearly polarized. The production and analysis of polarized light using such materials is shown in figure 4.10.

Birefringent materials possess one additional important property. Consider light containing both x-polarized and y-polarized components that strikes a slab of such a material at an angle of 90° to the surface. In this case neither polarization is bent so both travel along the same direction in the material, but at different velocities. One of the polarizations therefore gets ahead of the other, causing the phase difference between the two polarizations to change as the light travels

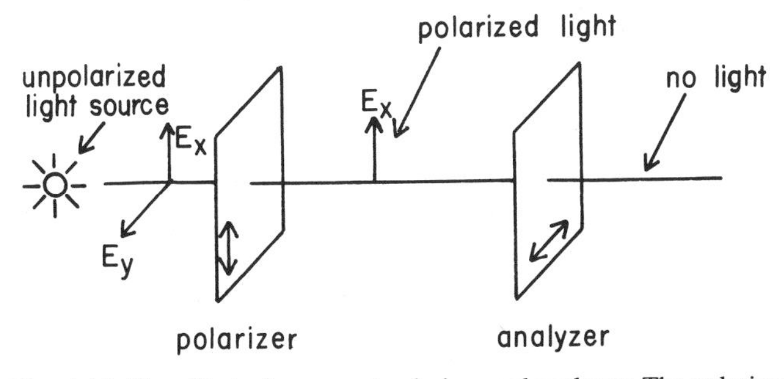

Fig. 4.10 The effect of a crossed polarizer and analyzer. The polarizer allows only light polarized along the x-axis to pass, while the analyzer allows only light polarized along the y-axis to pass. Since there is no light polarized along the y-axis emerging from the polarizer, no light passes through the analyzer.

through the material. If the two polarizations started out in phase (zero phase difference), they will emerge with a phase difference that depends on the thickness of the material. Devices that utilize this effect are called *phase retarders* and are extremely useful in many optical applications. For example, if the thickness is carefully adjusted to cause a 90° change in the phase difference, light that is linearly polarized upon entering the material emerges as circularly polarized light. Likewise, circularly polarized light incident on this slab emerges as linearly polarized light. Obviously, varying the thickness of the material produces many other possibilities.

Birefringence in Nematic Liquid Crystals

The anisotropy of nematic liquid crystals causes light polarized along the director to propagate at a different velocity than light polarized perpendicular to the director. Nematic liquid crystals are therefore birefringent. This fact is never more obvious than when a liquid crystal is placed between crossed polarizers. Normally no light emerges from crossed polarizers, because the light emerging from the first polarizer is completely absorbed by the second polarizer. Insertion of an isotropic material does not change this because the polarization of the light is unchanged as it travels through an isotropic material. Now consider the case where the polarized light from the first polarizer is oriented such that the direction of polarization makes an angle other than 0° or 90° with the director of the liquid crystal. From our past discussion, we know we can consider this light to be composed of light polarized along the director and light polarized perpendicular to the director with zero phase difference (figure 4.4). In passing through the liquid crystal, the two polarizations get out of phase and in general emerge as elliptically polarized light. Since the electric field of elliptically polarized light is constantly rotating completely around during each cycle, it is parallel to the polarization axis of the second polarizer twice during each cycle. Some light will therefore emerge from the second polarizer. Introduction of a liquid crystal between crossed polarizers in general causes the field of view to appear bright, whereas with no liquid crystal between the polarizers it was dark. It was this effect that was discovered by Otto Lehmann and led him to the correct conclusion that liquid crystals must be anisotropic.

Although liquid crystals normally appear bright when viewed between crossed polarizers, there are two conditions under which they continue to appear black. If the polarized light incident on the liquid crystal has its polarization direction either parallel or perpendicular to the director, all the light is polarized along this one direction in the liquid crystal and there is no need to consider light polarized at 90° to this direction. Since only one polarization is present, it propagates through the liquid crystal at one velocity and emerges polarized along the same direction. It is therefore extinguished by the second polarizer.

Pictures taken of liquid crystals under the microscope are usually taken with the sample between crossed polarizers. The director usually points in different directions at different points within the sample. Areas where the director is oriented parallel or perpendicular to the axes of the polarizers appear dark, while areas where the director makes an angle with the polarizer axes other than 0° or 90° appear bright. This is best illustrated in plate 1. Examination of this plate and others reveals an additional fact: there are many places where the brightness changes abruptly, indicating that the orientation of the director must also change abruptly there. These lines are called *disclinations* and represent places where the director is really undefined, since it points in many directions within an extremely small region.

These disclinations are therefore defects and are interesting in their own right, as will be seen in a later chapter.

Plate 1 reveals that point disclinations also exist. At many places in the picture four dark areas converge at right angles. Since dark areas occur when the director is along one of the polarizer axes, these four dark areas must define the two axes of the crossed polarizers. One director configuration that has four areas where the director is parallel to the axes of the crossed polarizers is where the director always points away from a point. This is shown schematically in figure 4.11. Although the director points away from the center of each of the crosses in plate 1, the orientation of the dark areas is determined by the crossed polarizers. That is why the dark areas at the point of convergence are all oriented in the same direction. These convergence points in the picture represent defects embedded in regions where the director is otherwise well defined.

Because the degree of orientational order in a liquid crystal varies with temperature, the indices of refraction for light polarized parallel and perpendicular to the director also change with temperature. At higher temperature where the order parameter is smaller, the two indices are closer together in value. The opposite is true at lower temperature in the liquid crystal phase where the order parameter is

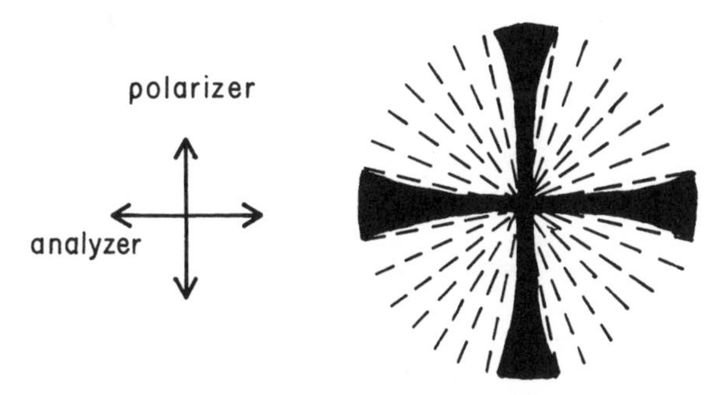

Fig. 4.11 Bright and dark areas around a defect in a liquid crystal viewed with the sample between a crossed polarizer and analyzer. The director always points away from the defect point. No light emerges from the analyzer when the director is parallel to either the polarizer or analyzer axis.

larger. The data for a representative liquid crystal are shown in figure 4.12. Notice that at the transition to the isotropic liquid the anisotropy disappears and one index of refraction serves for all polarizations.

THE SCATTERING OF LIGHT BY LIQUID CRYSTALS

One of the most obvious characteristics of liquid crystals is that they appear cloudy in bulk quantities, whereas isotropic liquids appear clear. The reason for this lies again with the fact that light is an electromagnetic wave and therefore interacts with matter as it propagates.

We saw before that light propagates only in the forward direction as it travels inside a material. Only at boundaries between two different materials can its path be changed. This is true provided there are no changes going on in the electric or magnetic properties of the material. For example, if the electric susceptibility at some point in a substance is changing in time as an electromagnetic wave passes, the polarization produced by the wave is not as simple as before. This

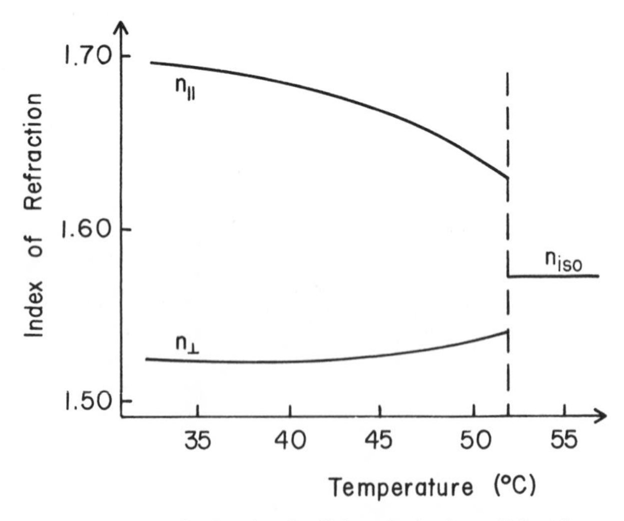

Fig. 4.12 Indices of refraction for light polarized parallel and perpendicular to the director of a typical liquid crystal. The vertical line denotes the phase transition from liquid crystal to isotropic liquid.

more complicated response plus the original wave can cause a new electromagnetic wave to propagate in a direction other than the forward direction. This produces a *scattered* wave and is only possible as long as these changes are taking place in the substance.

To illustrate this mechanism, consider the case of an air gun shooting a stream of ping pong balls at a stationary ping pong paddle. After striking the paddle, each ping pong ball is reflected in the same direction, producing a stream of balls representing the reflected wave. Obviously, in this case there is no transmitted (refracted) wave. Notice, however, that only one reflected wave is possible. Now imagine moving the paddle randomly as the stream of balls strikes the paddle. This causes the velocity of the paddle to change with time and you can imagine that the result is streams of ping pong balls reflected in many directions. These various streams represent scattered waves, and only exist as long as the orientation of the paddle varies in time.

Because the director of a liquid crystal responds so sensitively to changes of any kind, its orientation tends to fluctuate slightly simply due to the nearly random motion of the molecules. This causes the orientation of electric polarization to fluctuate, which scatters light incident on the liquid crystal. It is this scattered light that makes the liquid crystal appear cloudy. Isotropic liquids do not scatter light because the lack of a preferred direction for the molecules means there is no anisotropy in the electric polarization and therefore nothing that can fluctuate.

Smectic liquid crystals possess an additional mechanism by which they scatter light. The layer structure of smectics means that the electric and magnetic properties of the liquid crystal also depend on position, being different within a smectic layer than between two layers. The same nearly random motion of the molecules causes the layers to both undulate and vary in thickness slightly in a fluctuating fashion. In smectic liquid crystals, therefore, light is scattered by fluctuations in both the director and the layer structure.

CHAPTER 5

Light and X-Rays in Other Types of Liquid Crystals

The twisted structure of chiral nematic liquid crystals and the layered structure of smectic liquid crystals affect the propagation of light in ways quite different from what occurs in nematic liquid crystals. We shall see in this chapter that strikingly beautiful effects occur in chiral nematic liquid crystals when the pitch of the helical structure is about equal to the wavelength of light in the liquid crystal. This phenomenon forms the basis for a number of useful applications. Likewise, similar effects occur in smectic liquid crystals for electromagnetic waves with a wavelength about equal to the layer thickness. This thickness is significantly smaller than the wavelength of visible light, but falls within the X-ray part of the electromagnetic spectrum.

CIRCULAR BIREFRINGENCE

As discussed in chapter 1, the director in a chiral nematic liquid crystal is not constant, but rotates about an axis in a helical fashion. This additional structure interacts with light propagating through the liquid crystal in a way very different from nematics, producing some of the most interesting optical effects in liquid crystals.

Let us assume the director rotates about the z-axis. This means that at some points in the liquid crystal the director points along the x-axis, and at other points it is aligned with the y-axis. In fact, it rotates very uniformly through all directions in the XY plane. This situation is illustrated in figure 5.1, where the director is shown as an arrow pointing in different directions along the z-axis. If you follow the head of this arrow as it moves along the z-axis, it outlines a helix that is shown in the figure as a dashed line. This helix is easily visualized if you imagine being a flea walking along the threads of a screw. If you walk in the same direction along a single thread, your path will

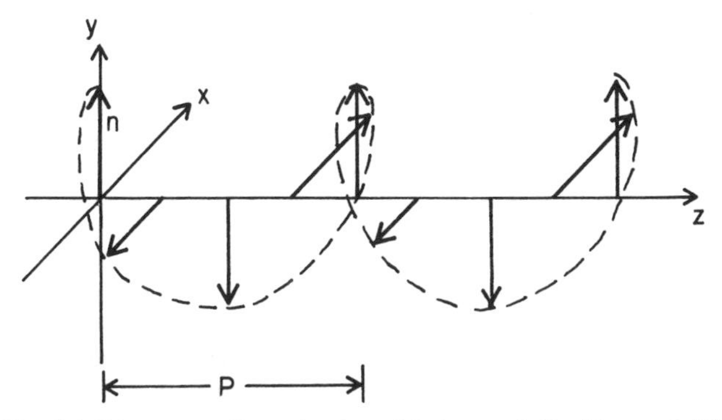

Fig. 5.1 Director configuration in a chiral nematic liquid crystal. The arrows represent the director, which is oriented differently for regions of liquid crystal along the *z*-axis. *P* is the pitch of the structure and the dashed line is the helical path of the tip of the arrow. The helix is right-handed (see text).

be the same as the helix drawn in figure 5.1. Perhaps the most important property of any helix is the distance along the *z*-axis for one full turn of the helix. As mentioned in chapter 1, this is called the pitch of the chiral nematic liquid crystal and is labeled in the figure as *P*. In our analogy of the flea walking along a single thread of a screw, the pitch would be the distance from one thread to the next measured parallel to the long axis of the screw.

Consider light linearly polarized along the *x*-axis propagating along the *z*-axis of a chiral nematic liquid crystal. This light must interact with parts of the liquid crystal having a director oriented at all angles with respect to the polarization of the light. Now consider what happens to light linearly polarized along the *y*-axis. It too must interact with parts of the liquid crystal having a director oriented at all angles relative to the polarization axis. In fact, light linearly polarized along any direction in the *XY* plane must interact with parts of the liquid crystal having a director oriented at all angles relative to the polarization axis. This means that linearly polarized light propagates along the *z*-axis with the same velocity regardless of its axis of polarization. Since components of light polarized along both the *x*-axis and *y*-axis travel together, no change in the phase difference between these two components results. This is different from nematic liquid crystals where a change in the phase difference between the components polarized along the x-axis and y-axis normally occurs.

The situation is quite different for circularly polarized light. Just as circularly polarized light can be either left-handed or right-handed, the structure of a chiral nematic liquid crystal can be either right-handed or left-handed. The definition, however, is slightly different from the definition used for circularly polarized light. Imagine looking back along the z-axis, observing the director for parts of the liquid crystal that are closer and closer to you. The director appears to rotate either counterclockwise or clockwise. If it seems to rotate counterclockwise we call it a right-handed helix; if it appears to rotate clockwise, it is a left-handed helix. Returning to our analogy of the thread of a screw, imagine looking along the axis of a screw as a nut is rotated so it approaches your eye. Is it rotating clockwise or counterclockwise? For most screws, the nut rotates counterclockwise, indicating that most screws are right-handed. Notice that it makes no difference which end of a screw you use to start the nut. A right-handed screw is simply right-handed, meaning that the nut must rotate counterclockwise if it is to move closer to your eye.

Now consider what happens when circular polarized light passes through a chiral nematic liquid crystal. There are two possibilities: either the light and the liquid crystal have the same sense of chirality (right-handed or left-handed) or they have the opposite sense of chirality. The interactions of the light with the chiral structures are different in these two cases, and this gives rise to two different propagation velocities. It is like trying to twist a nut onto a screw. Once again there are two situations. Either the nut and screw have the same chirality or opposite chirality, and the success you have in getting the nut to twist onto the screw is very different for the two cases. Because of the two different velocities, we can assign an index of refraction for right circularly polarized light, n_R, and another index of refraction for left circularly polarized light, n_L. The values of these two indices are equal for nematic liquid crystals, but unequal for chiral nematic liquid crystals. We could describe this situation by saying that chiral nematic liquid crystals possess *circular birefringence* in much the same way we said nematic liquid crystals possess linear birefringence.

In general, this circular birefringence depends on the amount of orientational order, so it too is greatest at lower temperatures and least at higher temperatures. Even more important, however, is the fact that the difference between n_R and n_L varies with wavelength in a much more dramatic fashion than the linear birefringence of nematics does. The reason for this will be clear later.

Optical Activity

In analyzing what happens to light as it propagates through a nematic liquid crystal, we considered the incoming light to be made up of light polarized along the *x*-axis and *y*-axis and then followed what happened to each of them. By adding the two together after they propagated through the liquid crystal, we were able to determine the polarization of the light as it emerged. For chiral nematic liquid crystals, we must do the same thing, but instead of thinking about the incoming light as made up of two linear polarizations, we must imagine the incoming light as composed of the two possible circular polarizations, right and left.

Let us consider the simple case of equal amounts of right and left circularly polarized light traveling down the *z*-axis. For the moment, assume no liquid crystal is present so that the two polarizations travel with the same velocity. An example of this situation is given in figure 5.2, where the view is for an observer situated at some point along the *z*-axis looking backward along the *z*-axis as the light passes. From this vantage point, the electric field rotates clockwise for the right circularly polarized light and counterclockwise for the left circularly polarized light as it passes. This is shown in figure 5.2 by labeling the positions of the electric field for the two polarizations with numbers indicating the sequence in time. Notice that at every time in the sequence the two electric fields are directed equally on either side of the *y*-axis. There is therefore no net electric field along the *x*-axis, so the total electric field must always point along the *y*-axis. The light must be linearly polarized along the *y*-axis. Equal amounts of right and left circularly polarized light therefore produce linearly polarized light in a medium with no circular birefringence.

What happens as this light that is linearly polarized along the *y*-axis enters a chiral nematic liquid crystal? The electric fields for the right and left circular polarizations continue to rotate, but because

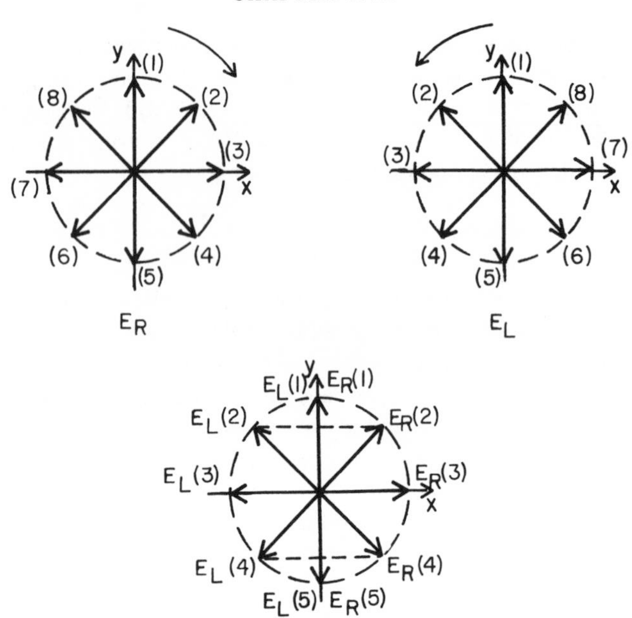

Fig. 5.2 Addition of right- and left-handed circularly polarized light. The upper two diagrams show the electric field direction at eight times for each of the circular polarizations. The numbers give the times in chronological order. The lower diagram shows how these combine to produce an electric field that always points along the *y*-axis.

they travel with different velocities, one gets ahead of the other. Let us assume the left circularly polarized light travels slower than the right, so after traveling through a chiral nematic liquid crystal of thickness *d*, the right circularly polarized light has gotten ahead of the left circularly polarized light. What emerges at any time, therefore, is right circularly polarized light combined with left circularly polarized light that entered the liquid crystal before it. Thus the right circularly polarized light had more time to rotate than the left circularly polarized light, so it is at a more advanced angle. This is shown in figure 5.3 by assuming the left circular polarization emerges along the *y*-axis, with the right circular polarization ahead of the *y*-axis by an angle θ. As the light continues to pass by this point, the electric fields for the two polarizations continue to rotate, but now in exactly opposite fashion. As can be seen in figure 5.3, the two electric fields now fall at every time in the sequence on opposite sides of an axis inclined relative to the *y*-axis. The light must therefore be linearly polarized along this direction. The chiral nematic liquid crystal has caused the axis of linear polarization to rotate clockwise as seen by an observer looking at the light as it approaches. Since the angle through which the polarization is rotated depends on how far behind the left polarization gets relative to the right polarization, this angle must be proportional to the thickness of the chiral nematic liquid crystal. The value of the angle divided by the thickness of the sample therefore does not depend on the thickness of the sample. We define

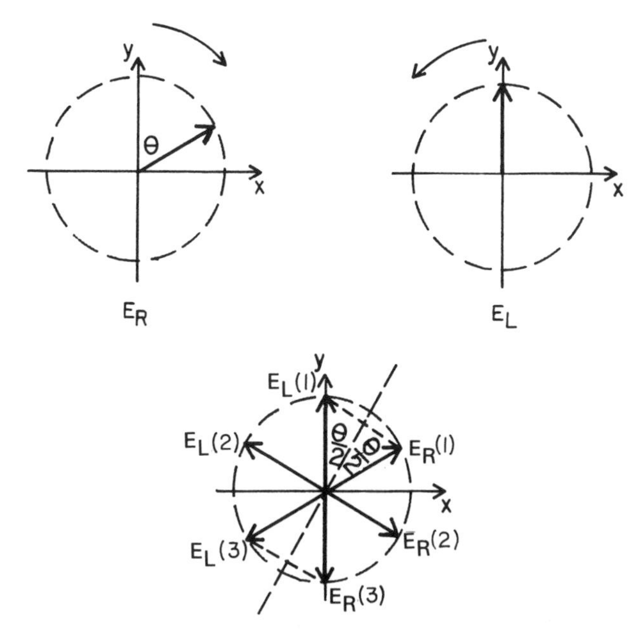

Fig. 5.3 Addition of right- and left-handed circularly polarized light when the electric field for the right-handed polarization is ahead of the left-handed polarization by an angle θ. The upper two diagrams show the two polarizations, while the lower diagram reveals how they combine to produce an electric field that always points along a line which makes an angle of θ/2 with the *y*-axis.

this ratio as the *optical activity* of the chiral nematic liquid crystal. Chiral nematics typically have quite a high amount of optical activity, with values of 300°/mm quite standard.

If instead of the situation just described, we imagine the case where the right circular polarization travels at a slower velocity than the left circular polarization, it is not difficult to see that the linear polarization axis will be rotated in a counterclockwise fashion for an observer watching the light emerge from the liquid crystal. Counter-clockwise rotation of the axis of polarization is assigned a negative value of optical activity, while clockwise rotation of the axis of polarization is assigned as positive value. Positive values for the optical activity result when the left circular polarization travels slower than the right, while the opposite situation leads to negative values for the optical activity.

Constructive Interference

Previously I discussed how electromagnetic waves are emitted by the charges in a material as an electromagnetic wave passes through it. Because all parts of the material are emitting waves, they tend to cancel except for waves traveling in the same direction as the original wave. This is true unless the electric or magnetic properties of the material themselves vary from one place to another, since in this case the waves being emitted from different parts of the sample are not all the same. A particularly fascinating case of this is when the electric or magnetic properties of the material repeat themselves over a distance equal to half the wavelength of the light. As shown in figure 5.4, this leads to the situation where the waves emitted by identical regions of the material add together in the backward direction.

The fact that the identical regions are separated by a distance equal to half the wavelength causes the electric field of each wave traveling backward to be in phase as it emerges from the sample. The waves therefore add together instead of canceling out. This phenomenon is called *constructive interference* and the result is that a sizable electromagnetic wave is not only transmitted, but also reflected.

Anyone who has watched a marching band perform various routines can understand why constructive interference occurs. Imagine a band marching along in rows of people separated by four paces. Be-

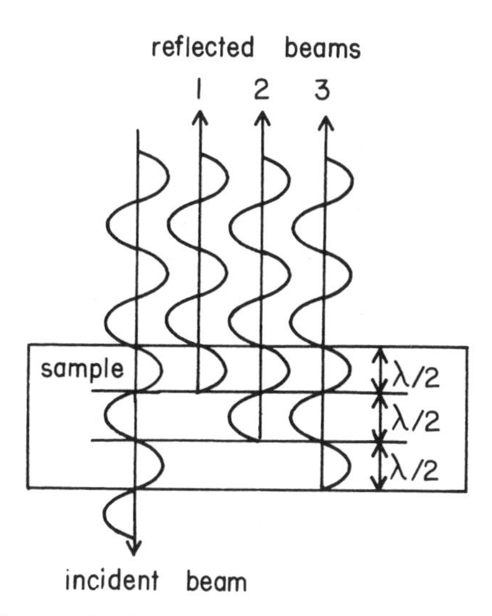

Fig. 5.4 Constructive interference when identical regions in the sample are separated by $\lambda/2$. All of the reflected light is in phase, thus generating a strong reflected beam.

cause the formation repeats itself over a distance of four paces, we can make the analogy that the band represents a wave with a wavelength equal to four paces. At some point the first person in the first column reverses his or her direction (stepping to the side slightly to avoid a collision) and the following people in the first column do the same thing at the same place. If the people in the second column perform the same maneuver at a point two paces farther along than the first column, the two columns will still be arranged in rows as they march in the opposite direction. The other columns can perform the maneuver at points separated by two paces, with the result that the entire band will end up marching in the opposite direction but still arranged in rows. The band members in each row have changed, but arrangement of members into rows with four paces in between is preserved. Just as in the case of electromagnetic waves, the band columns started out in phase with each other and ended up in phase with each other. This would not have been the case if the columns had performed their maneuvers at points separated by one or three paces.

CHAPTER FIVE

SELECTIVE REFLECTION

Constructive interference occurs in chiral nematic liquid crystals when the pitch of the helical structure is equal to the wavelength of light in the liquid crystal, since the helical structure repeats itself over a distance equal to half the pitch. If light with various wavelengths (white light, for example) is incident on a chiral nematic liquid crystal, most of it will be transmitted with some optical activity—except for light with a wavelength in the liquid crystal equal to the pitch. This phenomenon is called *selective reflection*, because only one wavelength is reflected. If this wavelength falls in the visible range, the light will possess a certain color. For this reason chiral nematic liquid crystals often appear brightly colored in reflection, with the color determined solely by the pitch of the liquid crystal.

Imagine an experiment where equal amounts of right and left circularly polarized light are incident on a chiral nematic liquid crystal. If the colored light being reflected is examined, it is found that it is either right or left circularly polarized, depending on whether the helix of the chiral nematic is right- or left-handed. The fact that the repeating structure is itself right- or left-handed, produces the additional effect that the constructive interference works for one circular polarization but not the other. Figure 5.5 shows some data for an

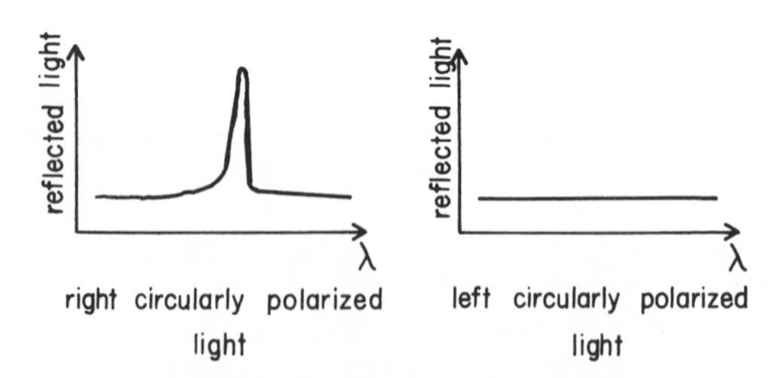

Fig. 5.5 Typical results of an experiment that measures the light reflected from a chiral nematic liquid crystal. Only one circular polarization is reflected, and only when the wavelength of the light in the liquid crystal equals the pitch.

experiment similar to the one imagined. Notice that a very narrow range of wavelength is reflected and only for one polarization of light.

One interesting example of selective reflection in nature is the color of some beetles. At some stage in the development of the cuticle of these beetles, a liquid crystalline substance is secreted. This substance then hardens, with the molecules fixed in place with the orientational order of the chiral nematic phase. The pitch of the helix determines what color of light is selectively reflected, and as expected, the light reflected from these beetles is circularly polarized.

Anomalous Optical Activity

The reflection of one circular polarization also has a dramatic effect on the optical activity. Of course, for thick enough samples and the proper wavelength of light only one circular polarization emerges from the liquid crystal since the other is completely reflected. In this case the optical activity cannot be measured. For wavelengths near this value, however, the index of refraction for the polarization being reflected changes abruptly from a value much smaller than the index for the opposite polarization to a value much larger than the other index. Since the sign of the optical activity depends on which index is larger, this means that the optical activity abruptly changes from one sign to another at this wavelength. In addition, the size of the optical activity depends on the difference between the two indices of refraction, so this change in sign is actually from a very large value of the optical activity of one sign to a very large value of the other sign. Data for a typical liquid crystal is shown in figure 5.6, clearly showing the dramatic effect for light of a particular wavelength. This anomalous behavior can produce extremely large rotations of the axis of linear polarization using very thin samples. For example, rotations of 10,000°/mm are quite common.

Chiral Nematics between Crossed Polarizers

Nematic liquid crystals appear bright between crossed polarizers due to the fact that the linear birefringence changes the linearly po-

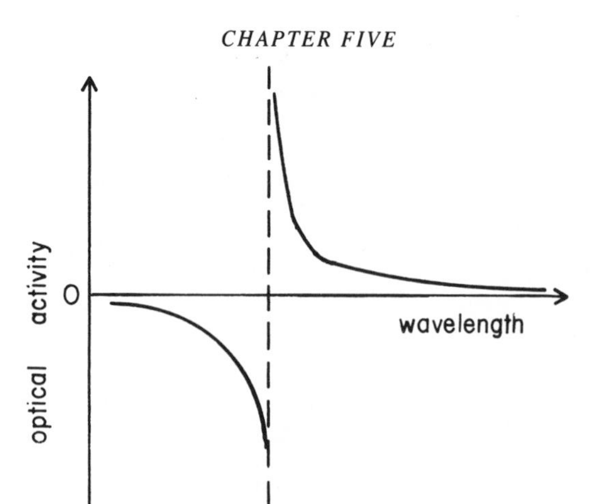

Fig. 5.6 Anomalous optical activity in chiral nematic liquid crystals. The optical activity changes abruptly when the wavelength of the light in the liquid crystal equals the pitch.

larized light to elliptically polarized light, some of which is transmitted by the second polarizer. Since the amount of linear birefringence is somewhat wavelength dependent, more light at one end of the spectrum is transmitted relative to light at the other end of the spectrum. It is therefore possible for nematic liquid crystals to appear slightly colored when viewed between crossed polarizers.

The situation is quite different for chiral nematic liquid crystals. The first polarizer ensures that linear polarized light enters the sample. The optical activity of the chiral nematic, however, rotates the axis of polarization so that some light can get through the second polarizer. The more the rotation of the axis of polarization, the brighter the sample appears. But we have just seen that wavelengths of light in the sample that nearly equal the pitch are rotated the most. This light will be transmitted through the second polarizer much more than other wavelengths, producing a specific color when viewed by an observer. The color depends only on the pitch of the chiral nematic liquid crystal, thus making it quite easy to determine the pitch of a sample placed between crossed polarizers. This effect can produce bright colors, as shown in plate 2.

Constructive Interference in General

Constructive interference is also possible even if the light is propagating at an angle to the *z*-axis. This can be seen in figure 5.7, where the waves emitted from two identical points in the liquid crystal have been drawn emerging as though they were reflected by the repeating planes in the chiral nematic. To see under what conditions constructive interference occurs, we must calculate how much farther the light reflecting off the lower plane travels relative to light reflecting off the upper plane. If this distance equals one wavelength, then the emerging reflections will be in phase and combine to give constructive interference.

Notice in figure 5.7 that the two paths are equal after point *e* on the upper path and point *c* on the lower path. The extra distance light traveling along the lower path encounters is thus equal to distance (*ab* + *bc*) minus distance ae. To determine this extra distance, we extend the lower exit path backward and drop a perpendicular to this extension from point *a*. The two distances *ab* and *bd* are equal. This means that the extra path distance (*ab* + *bc* − *ae*) equals (*cd* − *ae*), which is just the distance *df*. But *df* is a side of the right triangle *adf*, which

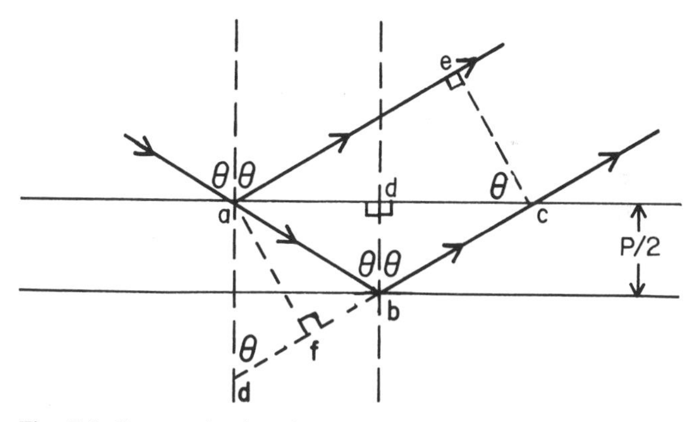

Fig. 5.7 Constructive interference in a chiral nematic liquid crystal when the light is incident at an angle θ with the perpendicular to the identical planes. The lower path is longer than the upper path by $P\cos\theta$ (see text), so constructive interference occurs when the wavelength of the light in the liquid crystal equals $P\cos\theta$.

has the hypotenuse *ad* (equal to *P*) and an angle θ. Therefore, *df* is just equal to $P\cos\theta$. If the wavelength of the light in the liquid crystal is equal to $P\cos\theta$, then constructive interference occurs for light striking the perpendicular to the surface at an angle θ. Notice that this result gives the same answer as before if the light is propagating along the *z*-axis, since in this case $\theta = 0$ and the condition for reflection simply reduces to the wavelength equaling the pitch.

When chiral nematic liquid crystals are illuminated with white light, it is possible to observe many colors coming from the sample depending on the angle at which the observer views the sample. The longest wavelength reflected comes from the condition that the wavelength in the liquid crystal is equal to the pitch and the reflected light emerges parallel to the *z*-axis. The wavelength reflected at all other angles is less than this wavelength, with the wavelength decreasing as the angle with the *z*-axis increases. It is therefore possible for chiral nematic liquid crystals to appear red from one direction, but then pass through yellow, green, etc., as the sample is viewed at larger and larger angles.

It is also possible to observe colors in a chiral nematic liquid crystal even if the pitch is much longer than the wavelength of light. Viewing the liquid crystal at large angles is one way. A second way involves another aspect of constructive interference. In constructive interference, the requirement for a reflection is that the extra distance light from identical points in the sample must travel should equal the wavelength of the light in the sample. This ensures that all the reflected waves would be in phase and constructively add together to form a strong wave. Obviously, the same result is possible if the extra distance is equal to two wavelengths, three wavelengths, etc. In other words, the condition for constructive interference is really $P\cos\theta = m\lambda$, where *m* is a positive integer and λ is the wavelength of light in the liquid crystal. The pitch therefore can be significantly greater than λ and this condition still can be met using a correspondingly larger value of the integer *m*. Many chiral nematic liquid crystals have a pitch near or slightly greater than the wavelength of light, so these liquid crystals are quite colorful. If the pitch is less than the wavelength of light in the liquid crystal, however, little color will be observed.

The Cano Wedge

From our discussion of selective reflection and anomalous optical activity, it is clear that the pitch of a chiral nematic liquid crystal can be measured by observing the wavelength at which selective reflection or anomalous optical activity occurs. One problem with both of these techniques is that the wavelength of light in a liquid crystal is different from the wavelength of the same light in air. The reason for this change is the decrease in the velocity of light when it enters the liquid crystal, a subject covered in chapter 4. Selective reflection or anomalous optical activity occurs when the wavelength of light in the liquid crystal equals the pitch of the chiral nematic. This same light has a longer wavelength in air (by a factor equal to the index of refraction), and this longer wavelength is what one measures in an experiment. Therefore, in order to measure the pitch, the index of refraction of the liquid crystal must be known. This requires an additional experiment.

Fortunately, another method for measuring the pitch exists, which does not require knowledge of the index of refraction. This method involves using two glass surfaces that continuously get farther and farther apart with a chiral nematic liquid crystal between the two surfaces. One can use two flat glass surfaces to form a wedge or a lens can simply be placed on top of a flat piece of glass. Either of these arrangements is called a *Cano wedge*, after the French scientist R. Cano, who first developed this technique in the 1960s. If the chiral nematic between the two surfaces orients with the pitch axis more or less perpendicular to the two glass surfaces, then there will be positions in the wedge where an integral number of half turns of the helix fit exactly between the two surfaces. This is shown schematically in figure 5.8. In these regions the pitch of the chiral nematic is equal to its normal value, so the liquid crystal is undistorted. On either side of these regions, the liquid crystal is distorted since the pitch must be more or less in order to fit an integral number of half turns between the two glass surfaces.

Even more important is the fact that roughly midway between these regions the number of half helical turns must change by one. This causes a defect there, which is visible as a sharp line under a polarizing microscope. One therefore sees equally spaced lines in the

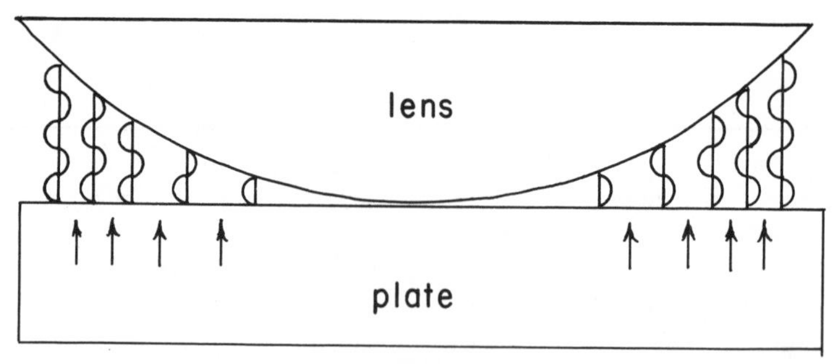

Fig. 5.8 A Cano wedge formed by a lens and flat piece of glass. A chiral nematic liquid crystal between the two glass surfaces is oriented with its helical axis perpendicular to the two surfaces (the curvature of the lens is exaggerated). The helix is shown for positions where an integral number of half helical turns fit between the surfaces. The vertical arrows show the locations of the defects, where the number of helical half turns changes by one.

case of a wedge or unequally spaced circles in the case of a lens on a flat surface. If the distance between these lines is measured, then knowledge of the wedge angle or the lens radius allows the difference in the distance between the two surfaces at each of the defects to be calculated. This must be equal to one half pitch.

The color due to selective reflection also changes between the lines, since the pitch slowly changes as the distance between the two surfaces increases or decreases. A beautiful photograph of a Cano wedge formed by using a lens is included as plate 8.

X-Rays and Smectic Liquid Crystals

X-rays are electromagnetic waves with a wavelength much shorter than visible light. Typically, X-rays have a wavelength equal to about 0.1 nanometer. Although chiral nematic liquid crystals do not reflect X-rays because the pitch is greater than the wavelength of the X-rays, constructive interference of X-rays does occur in smectic liquid crystals.

Smectic liquid crystals possess a layer structure, so here again there exist regions where the electric and magnetic properties of the liquid crystal are the same. In smectics, the distance between these regions is equal to the separation between layers. This distance is much smaller than the pitch of a chiral nematic, because the layer spacing in smectics is roughly equal to the length of the liquid crystal molecule. Molecular lengths are typically around 3 nanometers.

Figure 5.9 illustrates constructive interference from the layer structure of smectic liquid crystals. Notice how similar this diagram is to figure 5.7. We need perform no additional analysis, since all results remain the same. The condition for reflected X-rays is therefore $2D\cos\theta = m\lambda$ where D is the distance between layers. The factor of two results from the fact that the distance between identical regions is D in smectics, whereas it was $P/2$ in chiral nematics. For X-rays, the angle is usually measured from the X-ray beam to the planes in the smectic liquid crystal. This angle has been labeled ϕ in figure 5.9. Since θ and ϕ are complementary, $\cos\theta = \sin\phi$, so the condition for constructive interference for X-rays is usually written $2D\sin\phi = m\lambda$. This is a famous relationship, called *Bragg's Law*, which was first formulated to explain the reflection of X-rays from solid crystals.

X-ray experiments are an excellent way to measure the layer spacings in smectic liquid crystals. Let us calculate what the angle typically is for such an experiment. If D equals 3 nanometers and λ is equal to 0.1 nanometer, then the reflection corresponding to $m = 1$

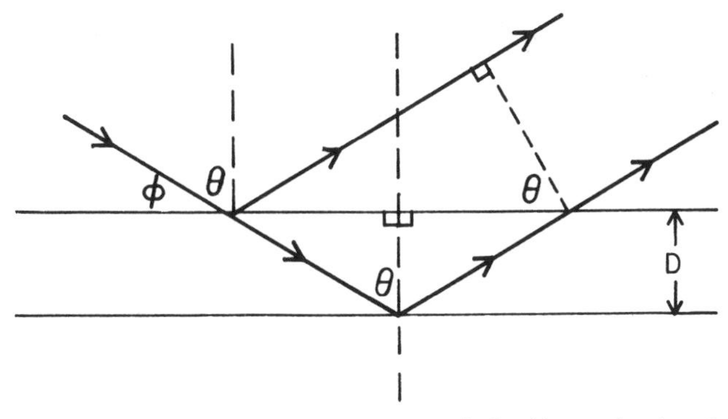

Fig. 5.9 Constructive interference in smectic liquid crystals when the *X*-rays are incident at an angle ϕ with the layers. The lower path is longer than the upper path by $2D\sin\phi$ (see text), so constructive interference occurs when the wavelength of the X-rays equals $2D\sin\phi$.

has an angle ϕ equal to 1°. These experiments must be designed to detect X-rays reflected at very small angles.

X-ray experiments on smectic liquid crystals have revealed that many different types exist. For example, some smectics have a layer spacing equal to roughly one and a half molecular lengths. These smectic phases occur for molecules that have a tendency to form molecular pairs in which the molecules align themselves parallel to each other by overlapping to some degree. This pairing is usually due to the existence of a permanent electric dipole parallel to the long axis of the molecule near one end. Figure 5.10(a) shows such pairs of molecules and the *smectic* A_d phase that can result. In some cases, such molecules form another type of smectic phase in which the molecules form layers without overlapping, with the electric dipoles concentrated in the boundary between every other layer. A diagram of this phase, called the *smectic* A_2 phase, is shown in figure 5.10(b). Notice that the structure repeats itself over a distance equal to two molecular lengths. This is quite different from the smectic *A* phase (called *smectic* A_1 with this nomenclature), in which the molecules within each layer point up or down randomly.

As mentioned in chapter 1, some smectic liquid crystals also possess an arrangement for the molecules within each layer. This produces a repeating structure of much smaller length (roughly a molecular width rather than a molecular length) along a direction parallel to the layers. It is possible to observe X-ray reflections due to these repeating regions and thereby gain knowledge about how the molecules are arranged. These experiments show that various arrangements exist, with a typical one having molecules located at the center and vertices of connected hexagons.

Chiral Smectic C Liquid Crystals

Usually when a liquid crystal possesses a chiral nematic phase, its smectic phases, if present, are either smectic *A* or smectic *C*. In some cases, however, the tendency for the molecule to form a twisted phase is so great that a smectic phase with a twist exists, either by itself or along with other smectic phases. One common twisted smectic phase is called the *chiral smectic C* phase or simply the *smectic C** phase. As can be seen in figure 5.11, each layer appears to be typical of a smectic *C* liquid crystal in that the molecules tend to

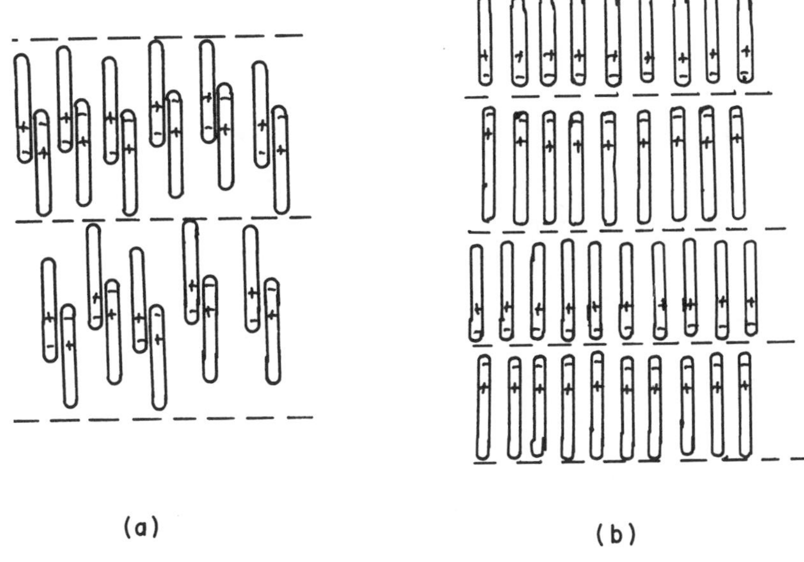

Fig. 5.10 The smectic phases formed by molecules that tend to form pairs. The spacing between the smectic layers is about one and a half molecular lengths in (a), while the distance between identical regions is two molecular lengths in (b). The + and − signs represent permanent electric dipoles at the end of each molecule.

orient themselves at an angle to the layers, but this direction rotates from one layer to the next. The angle between the director and the smectic layers remains constant, but the director outlines a cone as it rotates in going from layer to layer. This liquid crystal has a layer spacing about equal to the length of the molecules, but it also possesses a pitch (the distance over which the director makes a complete revolution about the cone). Chiral smectic *C* liquid crystals therefore reflect X-rays just as other smectic liquid crystals, but they can also reflect light, as is typical for chiral nematic liquid crystals. The pitch of these smectic *C** liquid crystals tends to be longer than the pitch of chiral nematics, but as explained earlier, they can still reflect visible light. Smectic *C** liquid crystals can also appear colored when viewed under a microscope for the same reason chiral nematics show colors.

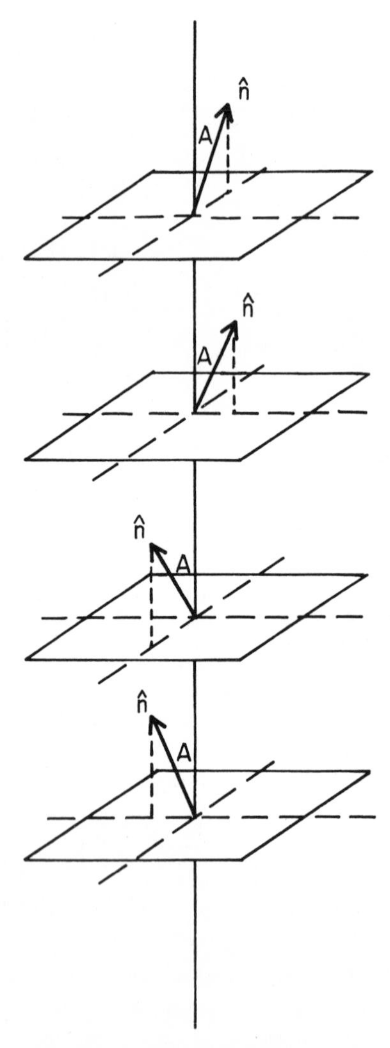

Fig. 5.11 Structure of the chiral smectic or smectic C^* phase. The planes represent the smectic layers. The director always makes the same angle with the smectic planes, but the orientation of the director rotates about the line perpendicular to the planes in going from one layer to the next.

Other Types of Smectic Liquid Crystals

As previously mentioned, unlike smectic *A* and smectic *C* liquid crystals where no positional order exists within the layers, there are smectic phases that display positional order within the layers. The molecules in a *smectic B* liquid crystal tend to be oriented perpendicular to the planes and are usually located at the corners and centers of a network of hexagons. The *smectic F* and *smectic I* phases are tilted analogues of the smectic *B* phase, in which the tilt of the molecules is perpendicular (smectic *F*) and parallel (smectic *I*) to the one side of the hexagons. Although there is little correlation between the hexagon networks from one smectic plane to another, the orientation of the hexagons in different planes is consistent. This type of order is called *bond orientational order*.

There are also smectic phases, both tilted and not tilted, which display strong correlations between the preferred positions of the molecules in the different smectic planes. These include the *smectic E* phase (not tilted and with rectangular packing), the *smectic G* phase (tilted and with rectangular packing), and the *smectic H* phase (tilted and with hexagonal packing).

There is also the *smectic D* phase, which seems to be some type of cubic arrangement of the molecules. There is little evidence for a layered structure, and the phase seems to be isotropic.

Just as the smectic *C* phase can exist in both an untwisted and twisted form, some of the smectic phases with positional order within the layers also have versions in which the direction of orientational order rotates in going from layer to layer. The *smectic H** and *smectic I** phases are the best examples of this.

Recently, a new type of chiral smectic *A* phase was reported. In this phase, normal regions with untwisted smectic *A* layers are separated by defect structures in which the direction of the layers changes abruptly. In going from region to region in a direction perpendicular to the director, the director rotates in a helix (although not continuously). Understanding defects in liquid crystals such as these is important, as we shall see in chapter 10.

In short, there are many types of smectic phases, which differ in the amount and type of both positional order within the layers and correlation of the order between the layers. The existence of all these different phases offers scientists the chance to examine how cooperative systems behave under varying conditions. For this reason, theoretical and experimental work on these smectic phases is an active area of research.

Freestanding Liquid Crystal Films

It is possible to make films using smectic liquid crystals in much the same way soap films are made. A small amount of liquid crystal is placed around a hole in a glass slide and the side of another glass slide is then drawn across the hole. Some of the liquid crystal is carried across the hole as the second glass slide is drawn, creating a thin film across the hole. The amount of liquid crystal material placed around the hole and the rate at which the second glass slide is drawn determine the thickness of the liquid crystal film. Films consisting of only a few smectic layers can be made in this way. These films are too thin to create interference between the light reflected from the two surfaces, so they appear gray. Thicker films, however, do cause interference at certain visible wavelengths, so these can be highly colored. In general, uniform free-standing films with areas of one square centimeter and stable for long periods of time can routinely be fabricated.

Liquid crystal films are mostly used for scientific investigations. When only a few layers thick, they represent a system that is virtually two-dimensional in character. If our general understanding of phase transitions is valid, then phase transitions in two-dimensional systems should be quite different from phase transitions in three-dimensional systems. Results from experiments on liquid crystal films support this idea.

Imaging Liquid Crystal Molecules

Although the amount of evidence for the fact that liquid crystals involve the orientational order of linear molecules is virtually overwhelming, there is nothing quite like seeing a picture of real liquid crystal molecules showing clearly how they tend to point in the same direction. Likewise, a picture showing the layered arrangement of molecules in a smectic liquid crystal would be extremely convincing.

Until recently, such a thought was only a dream, but the development of the scanning tunneling microscope (STM) has changed that forever.

The STM works by measuring the electrical current that passes from a substrate to a tiny metal tip held about 1 nm above the substrate. The amount of current (or height of the tip above the substrate required to maintain a certain current) depends on the atom sitting right below the tip on the surface of the substrate. As the tip is moved across the substrate, the current (or height) is measured at very closely spaced points. By combining many scans, a two-dimensional image can be formed, where the brightness at any point corresponds to the amount of current (or height of tip) at that point. Such a technique is capable of creating images of individual atoms on the surface of the substrate.

Recently, researchers performed an experiment where they placed a small amount of liquid crystal on an extremely flat crystal of graphite. The metal tip of the STM was placed in the liquid crystal and lowered closer and closer to the graphite surface. If the tip was placed close enough to the graphite surface to create an image, but not so close that an image of the graphite surface resulted, then an image of the atoms of the liquid crystal molecules was obtained. These images clearly show that the molecules are oriented parallel to each other, and that the molecules in smectic liquid crystals form layers. As amazing as these images are, there is one drawback to the technique. The molecules being imaged are very close to the graphite substrate, and as previously mentioned, nearby solid surfaces tend to exert strong forces on liquid crystal molecules. As a result, the molecules in the STM images show much more order than occurs in a bulk sample, and also orient along directions defined by the atoms in the nearby graphite surface. In any case, these images are a fitting tribute to the numerous scientists who utilized only skilled judgment and intellectual creativity to form our present picture of the various liquid crystal phases.

CHAPTER 6

Liquid Crystal Displays

We live in an age of information. This statement usually refers to the vast amount of data available due to the recent explosion in computer technology. There is another aspect to this development, however, which concerns the transfer of this information from machine to human being. For example, most computers must be equipped with an alphanumeric and/or graphics display in order for the user to benefit from the capabilities of the computer. The dedicated microprocessors one finds in automobiles, audio equipment, and aircraft also manage data that must be displayed for the user. Small alphanumeric displays are usually used to convey this information. It is therefore not surprising that rapid development in display technology has accompanied the advances made in computer technology. Liquid crystal displays (LCDs) have played an important role in this development, and appear to be destined to play an even wider role in the future. But before I can discuss how liquid crystal displays function, a brief introduction to the general field of information displays must first be presented.

Information Displays

All information displays utilize the ability to control light in order to function. By controlling what parts of a display are bright and what parts are dark, information is passed to the user. The most simple example of how this can be done is the seven-segment numeric display. Light from the area of each of the seven segments is controlled independently, thereby creating the possibility of producing each of the ten digits. Such a display is shown in figure 6.1(a). All the letters of the alphabet plus the ten digits can be produced if the light from the area of each of fourteen segments is controlled independently. Such a display is shown in figure 6.1(b). If a more pleasing display is desired, a five-by-seven dot matrix can be used. Here light from thirty-five different areas is controlled independently, producing all the letters and digits with more detail, as shown in figure 6.1(c). Notice that this last display requires that considerably more areas be controlled. The epitome of flexibility is a large-scale display where the entire field of view is composed of small areas (called pixels), each of which can be independently controlled. Such a display might have 360 rows and 720 columns (in other words 259,200 pixels), with the capability of displaying both alphanumeric and graphical information.

Regardless of the complexity of the display, the basic working principle remains the control of light from small areas of the display. This can be done in two ways. Each area can be equipped with the ability to generate light. This is called an *active display*, and two good examples are the cathode-ray tube (CRT) and light-emitting diode (LED). The CRT screen contains phosphors that emit light when struck from behind by an electron beam. Different areas are made to produce light by hitting only these areas with the electron beam. In an LED display, each area contains a small light-emitting diode, which gives off light when electrical voltage is applied to it. By controlling which diodes receive the applied voltage, various letters and numbers can be created. The second type of display does not generate light, but controls the amount of light that passes through or is reflected by the display. This is called a *passive display*. In some passive displays the light is generated by a device behind or beside the actual display; in other cases, ambient light is used. The number of

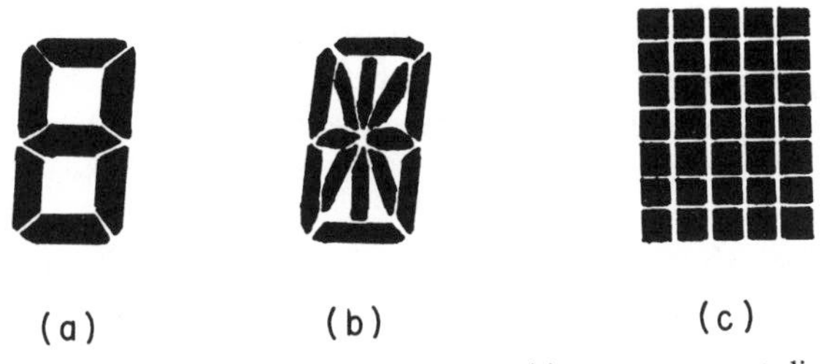

Fig. 6.1 Three different display patterns: (a) seven-segment display for numbers, (b) fourteen-segment display for numbers and letters, and (c) five-by-seven dot matrix display for numbers, letters, and other symbols.

types of passive displays is extremely limited, with LCDs being by far the most important.

Both active displays, and passive displays that are illuminated from behind or from the side, must use electrical power to generate light. Passive displays that utilize ambient light do not consume electrical power in order to generate light, and therefore require much less power to operate. This is the largest advantage of liquid crystal displays that rely on ambient light for their operation. The amount of power needed to operate these displays is much less than other displays, making LCDs ideal for battery operated equipment, such as wrist watches, portable radios and tape recorders, and pocket calculators.

With both active and passive displays, the control of light from each area is usually done by applying a voltage to a device in that area (the CRT is an obvious exception). How well the device functions as a display depends on how it responds to this applied voltage. For example, figure 6.2 illustrates how the light from a device might depend on the value of the applied voltage. There are two characteristics of the response that prove to be important. First of all, what value of the voltage is necessary to get a change in the brightness? This is called the *threshold voltage* and is indicated in the figure. The second characteristic is how much of a voltage increase is necessary to make the response of the device go from 10% brightness to 90% brightness? This is usually denoted by $V_{90} - V_{10}$; the smaller this quantity, the "sharper" the response. Although a "sharp" response (small $V_{90} - V_{10}$) is very advantageous in displays that are designed to have each area either bright or dark, there is another class of displays that utilize a *gray scale*. In this type of display, each area has the capability of producing various degrees of brightness. This can be accomplished by applying slightly different voltage values between V_{10} and V_{90}, and is obviously easier if the response of the device is not too "sharp."

Another characteristic of any display device is how quickly the device responds as the voltage is either applied or removed. An example of such a response is shown in figure 6.3. Notice that the response to applying a voltage versus removing the voltage may be different. The *switching time* T_{ON} is equal to the time between application of the voltage and a 10% response plus the time necessary for the device to go from a 10% response to a 90% response. Likewise, the switching time T_{OFF} is equal to the time between removal of the voltage and a 10% change plus the time necessary for a 10% to 90% change. In most liquid crystal displays, T_{ON} and T_{OFF} differ because the device turns on by responding to the application of a voltage, but simply relaxes back to its off condition when the voltage is removed. The characteristic of the liquid crystal that is most important in determining how quickly it responds is its *orientational viscosity*; that is, the amount of resistance in the liquid crystal when it is forced to change its director configuration. The time it takes the liquid crystal to relax after the applied voltage is removed strongly depends on its viscosity. Likewise, the time to switch on when the voltage is applied is also affected by the viscosity, but to a lesser degree.

Perhaps one of the most important characteristics of a display is the difference in brightness between an area turned on and an area turned off. In figures 6.2 and 6.3, the brightness in each state is indicated by the height of the curve above the voltage or time axis. This difference in brightness is loosely called the *contrast* of the device;

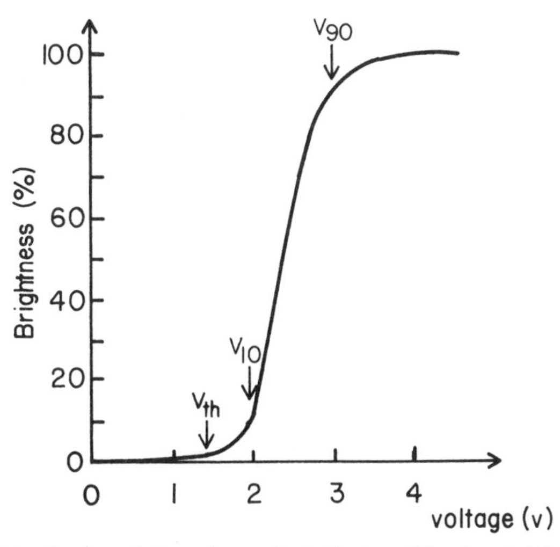

Fig. 6.2 Characteristics of a typical display. The threshold voltage V_{th}, 10% brightness voltage V_{10}, and 90% brightness voltage V_{90} are indicated.

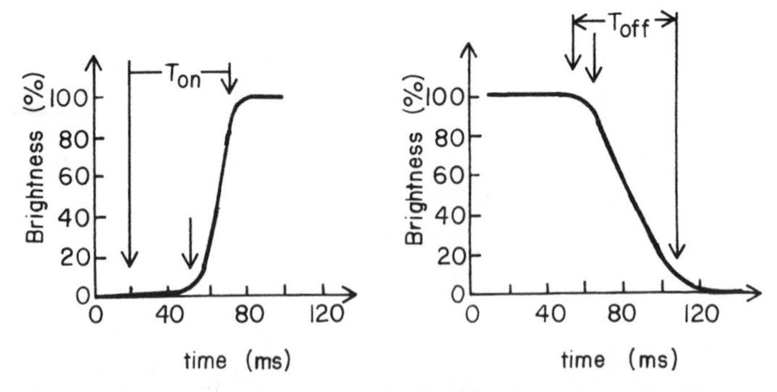

Fig. 6.3 Turn-on and Turn-off characteristics for a typical liquid crystal display. T_{on} and T_{off} are the times between application of the voltage change and a 90% change in brightness. The arrow in the middle shows the time for a 10% brightness change in each case.

specifically, it is defined as the difference in brightness divided by the larger of the two brightness values. Obviously, the closer the contrast is to one, the better. Another term often used in describing liquid crystal displays is *contrast ratio*. This is defined as the larger of the two brightness values divided by the smaller of the two values. Devices with good contrast have contrast ratios much greater than one; contrast ratios for all types of displays range from 10 to 40.

An interesting feature of some displays is that the threshold voltage and contrast ratio depend on the angle at which the display is viewed. This is especially true of passive displays, and we shall see that this is a disadvantage in many LCDs.

Addressing Displays

We must discuss how the voltages to the various devices that make up a display can be applied. If each area of a display is directly connected to circuitry that applies a voltage, we call this scheme *direct addressing*. The advantage is that each area is under constant control, but the disadvantage is that individual connections must be made to each area. Another method that can be used is called *multiplexing*. One example of this kind of addressing is a display where all the areas of a row are connected together and all the areas of a column are connected together. This configuration is shown in figure 6.4 for both

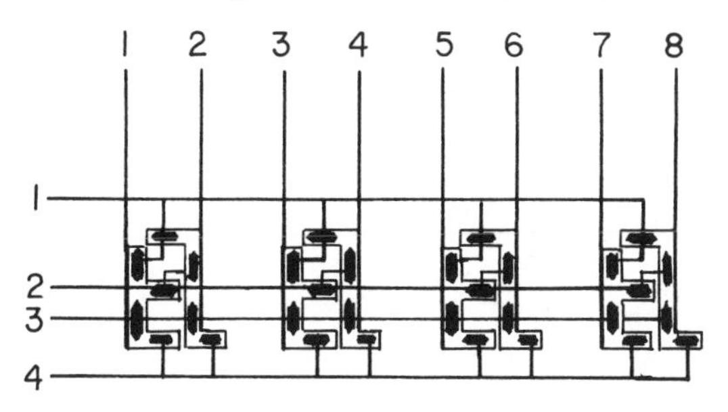

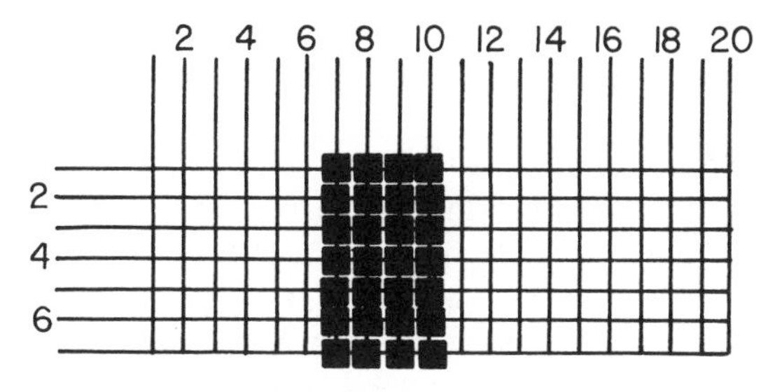

Fig. 6.4 Two types of multiplexed displays. In the top display, each row is connected to two segments in each digit and each column is connected to four segments of a single digit. Applications of a voltage to a row together with the desired voltages on each column turns on two segments in each digit as desired. The same procedure turns on the desired pixels in the matrix display. In a liquid crystal display, the rows are connected to an electrode on one surface while the columns are connected to an electrode on the other surface, with the liquid crystal between the two electrodes.

a seven-segment display and a large-scale graphics display. An individual pixel is turned on by making sure the proper row and proper column receive an appropriate voltage at the same time. Obviously, other pixels in the same row or same column receive some sort of applied voltage, but it must be low enough so as not to turn these pixels on. The obvious advantage of this arrangement is that far fewer connections to the display are required.

In addition, multiplexing does not allow you to apply the proper voltages to all areas at the same time. You might apply a voltage to row 1, while simultaneously applying the proper voltages to each of the columns. This would make all areas of the first row light up properly. Then you would apply a voltage to row 2 and simultaneously apply the proper voltages to each of the columns. This would make all areas of the second row light up correctly. In other words, you would have to cycle through all N rows one at a time, changing the column voltages each time. This means that any area that is on is only given the proper voltage once in each cycle of N rows, and receives smaller changing voltages during the rest of the time. Any area that is off receives a smaller changing voltage all of the time. The fraction of time an individual area receives the on voltage is called the *duty cycle* and is equal to $1/N$. Liquid crystal displays respond to the average voltage applied over a cycle. Making the applied voltage large when an area is to be turned on is one way to make the average voltage over a cycle for on areas larger than the off areas. Unfortunately, the larger N is, the smaller the fraction of time this larger voltage is applied and the smaller the difference in average voltage for on and off areas. There is therefore a limit on the number of rows, since the difference in average voltage (and thus the contrast of the display) decreases as the number of rows is increased. A "sharp" voltage response (small $V_{90} - V_{10}$) is the only remedy for this situation, since it might be possible to have the average voltage applied to off areas fall just below the threshold voltage, while the average voltage to the on areas (which is only slightly higher if N is large) falls somewhere around V_{90}. In this way good contrast can be achieved with a high *degree of multiplexing*, N.

One last comment should be made about multiplexing. It takes a small amount of time to apply the voltages to a single row and all the columns, so the time required for a full cycle through all the rows increases as N increases. If the duration becomes long enough, areas that are on will appear to flicker. It is therefore advantageous for multiplexed displays to have a somewhat long switching time when an area goes from on to off. This will keep the area on during part of the cycle and lessen the amount of flicker. Even better would be a display that remained on after the voltage is removed until it is turned off by another value of the applied voltage. Such a display is called a *storage display* and can be constructed using liquid crystals.

Dynamic Scattering Mode

The first liquid crystal displays utilized the light scattering properties of liquid crystal films when subjected to a large electric field. As discussed in chapter 3, this situation occurs when a liquid crystal that prefers to align with the director perpendicular to the electric field adopts a homeotropic texture between two pieces of glass. Application of a strong electric field causes charged impurities in the liquid crystal to flow, causing turbulence in the film. Although the unperturbed liquid crystal film is clear, the turbulent film intensely scatters light and therefore appears white. A reflector behind the display allows it to function using room light. Such a display is shown in figure 6.5. Evident from the figure is the fact that voltages in LCDs must often be applied to the cell by a *transparent electrode*. This is simply a layer of tin oxide or indium oxide coated on the glass. The layer is thick enough to act as an electrical conductor, but thin enough to allow most of the light to pass through. These transparent electrodes can be made in any size or shape, thereby creating the different areas for forming numbers, letters, or graphical information.

The voltage necessary to run a dynamic scattering mode LCD is about 10–20V, which is slightly higher than other LCDs. The liquid crystal must possess some charged impurities; this can be achieved by adding impurities to the liquid crystal, or in some cases the liquid crystal ordinarily contains sufficient charged impurities. In a dynamic scattering mode LCD, as in almost all LCDs, the voltages applied to the device are constantly reversed. This prevents both the buildup of charge on the electrodes and certain electrochemical reactions involv-

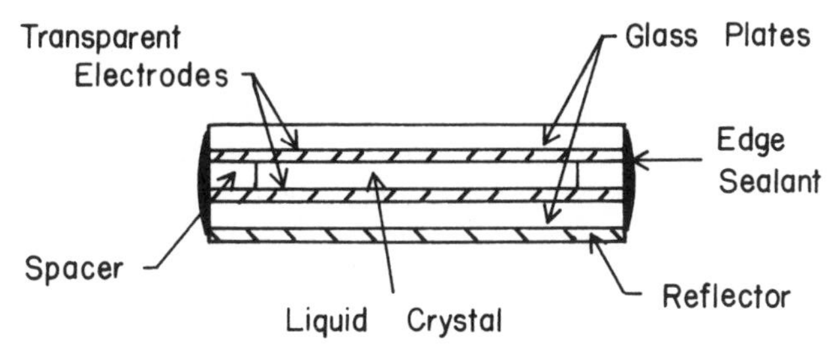

Fig. 6.5 Typical liquid crystal display cell. The reflector is only present if the cell is to be used in the reflection mode.

ing the liquid crystal. Since the dynamic scattering mode depends on the movement of charged carriers, the voltage cannot be reversed too many times each second, or else the charges will not have enough time between reversals to flow very far. Reversing rates up to about one thousand times per second work well.

The figures in a dynamic scattering mode LCD appear cloudy white on a clear background. Although colored filters can be added to the display, which improve the readability slightly, the contrast is still not very high. For this reason, dynamic scattering mode LCDs are seldom used anymore.

Chiral Nematic Mode

A storage display can be constructed by utilizing the dynamic scattering mode in a chiral nematic liquid crystal. If the molecules are forced to lie parallel to the glass surfaces, the liquid crystal adopts a texture with the helical axis perpendicular to the glass surfaces. In this state the cell appears clear. If the liquid crystal prefers to align the helical axis perpendicular to the electric field, the application of a fairly large voltage will cause the clear texture to appear cloudy, due to the fact that the texture breaks up into many small unaligned domains. This brightly scattering state remains even after the voltage is removed. An even higher and quickly reversing voltage restores the cell to its clear, aligned state. This type of LCD requires even higher voltages (40–100V to turn it on and nearly 300V to turn it off). The switching times are also fairly long (for display purposes), with $T_{ON} = 0.08$ second and $T_{OFF} = 1$ second. Like the dynamic scattering mode LCD, its contrast is not very good.

A different type of LCD can be constructed by using a chiral nematic liquid crystal composed of molecules that prefer to align parallel to an electric field. If the cell is correctly prepared, the liquid crystal adopts an unaligned texture (consisting of many small domains) that appears bright white. Application of a voltage forces the molecules to point perpendicular to the glass surfaces, producing a uniform texture that appears clear. There is no storage capability, however, as the liquid crystal relaxes back to the unaligned, bright texture as soon as the voltage is removed. Although the switching times are much shorter, the contrast is still quite poor.

Twisted Nematic Mode

The twisted nematic mode LCD was developed in the early 1970s and quickly became the workhorse of LCD technology. It remains the predominant type of LCD in use today, mainly because the last twenty years have brought continued improvement in its characteristics.

The *twisted nematic* (TN) display differs from the ones discussed so far by the addition of crossed polarizers to the outside of each piece of glass. As shown in figure 6.6, light incident on the cell from above is polarized along the right/left direction. In addition, the glass surfaces have been treated so the liquid crystal molecules prefer to lie parallel to the surface, but in the right or left direction for the top piece of glass and in or out of the page for the bottom piece of glass. The director of the nematic liquid crystal is therefore forced to twist through an angle of 90° within the cell. This twist is just like the twist of a chiral nematic liquid crystal, so the optical activity in the liquid crystal produces rotation of the polarization direction as light propagates through the cell. With the correct choice of cell thickness and liquid crystal material, the rotation of the polarization direction can be made to follow the twist of the director. The polarization direction of the light is therefore rotated 90° when it strikes the second polarizer. This polarizer allows light with an electric field directed into or out of the page to pass through, which is exactly the state of the light when it reaches the second polarizer. Behind the second polarizer is a reflector that causes the light to traverse back through the cell. Its polarization direction is still into and out of the page so it passes through the bottom polarizer, is rotated 90° by the liquid crystal again so it passes through the top polarizer, and then emerges from the cell. In this state the cell appears silvery clear.

The situation is quite different when voltage is applied to the cell. The nematic liquid crystal must prefer to align parallel to the electric field for the cell to work. If the applied voltage is above the threshold value for the Freedericksz transition (see chapter 3), the director configuration changes from one of twist to the deformed state shown in figure 6.6(b). Only a small amount of twist remains in the regions next to the glass surfaces. The polarization direction of light traversing the cell is rotated only slightly, meaning that almost all of the light that passes through the top polarizer cannot get through the bot-

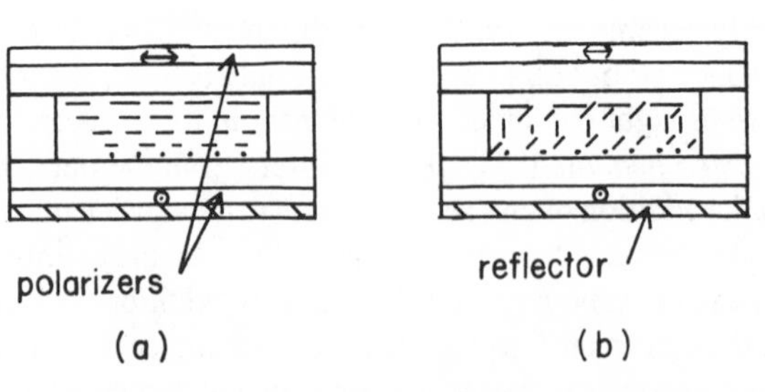

Fig. 6.6 Twisted nematic liquid crystal display. The polarizers on either side of the cell are crossed as shown. In the absence of an electric field (a), the surfaces induce a 90° twist of the director. This rotates the polarization axis of the light by 90°, thus allowing it to pass through the second polarizer. The presence of an electric field (b) removes this 90° rotation, therefore allowing no light to pass through the second polarizer.

tom polarizer. Since no light is reflected back out of the cell, areas with an applied voltage appear dark against those areas without an applied voltage that appear silvery clear. Removal of the voltage from any area causes immediate relaxation back to the 90° twisted texture. The result is a display with much better contrast and acceptable switching times.

A twisted nematic LCD can be operated in either of two modes. Figure 6.6 depicts a twisted nematic cell that is designed to utilize ambient light. The cell and reflector control the proportion of this ambient light that is redirected back to the viewer's eyes. A twisted nematic cell can also be designed to work in a transmissive mode. All that is necessary is to remove the reflector and place a source of light behind the cell. Light from the source is polarized by the rear polarizer, rotated 90° by the cell if no voltage is applied, and thus passes through the front polarizer. If a voltage is applied, however, the light polarized by the rear polarizer is not rotated by the cell and is therefore blocked by the front polarizer. The viewer sees dark areas against a bright background. Sometimes a transmissive LCD is made with the front and back polarizer parallel to each other. In this case little light gets through the cell if no voltage is applied, while much of the light is transmitted when an applied voltage is present. The viewer therefore sees bright characters on a dark background. Of course, when used in the transmissive mode, additional electrical power must be supplied to generate the light at the source.

One advantage of the twisted nematic LCD is that the threshold voltage is low, on the order of a few volts. However, its greatest advantage is its higher contrast. Unfortunately, this is achieved through the use of polarizers, which introduce problems of their own. For example, polarizers transmit between a third and half of the light incident on them. Using two polarizers therefore reduces the brightness of the display considerably. In addition, the optical activity of the cell and the polarizers themselves work differently depending on the angle at which the light is traversing the cell. Therefore, twisted nematic LCDs generally work best when viewed straight on, with serious degradation of performance as the viewing angle is changed. In addition, the "sharpness" of the brightness versus voltage curve is not great, meaning that a high degree of multiplexing becomes difficult with twisted nematic LCDs. Finally, the switching times for this type of LCD are not particularly fast (.02 to .05 second), so applications requiring a fast response are not feasible. These last two characteristics are the reason that liquid crystal television screens using the twisted nematic mode have not been widely developed.

One problem encountered early on with twisted nematic LCDs was that different areas of the cell would twist in opposite directions when no voltage was applied. This gives the cell a patchy appearance, and can significantly degrade its performance. The solution to this problem was to add a small amount of a chiral nematic liquid crystal to the mixture placed in the cell. A chiral nematic liquid crystal only twists in one direction (right- or left-handed), so it forces all parts of the cell to twist in the same direction.

In 1985 a slightly different version of the twisted nematic LCD was introduced. Instead of a 90° twist, the director of the liquid crystal in this new type of cell rotates 270° in going from one glass surface to the other. The principle of operation remains much the same, because a 270° rotation of the polarization direction of the light produces almost identical effects as a 90° rotation. The advantages of this *super twisted nematic* (STN) display are a slightly "sharper" brightness versus voltage characteristic and less of a problem when the viewing angle is changed. The disadvantage is that the higher degree of twist accentuates the fact that the display affects different wavelengths of light (colors) in slightly different ways, making it more difficult for the off state to appear completely dark. One recent attempt to rectify this last problem is to prepare *double super twisted* (DST) cells, with one cell right on top of the other. The twist of the two cells is opposite and only one of the cells is switched on and off by an applied voltage. Since the unswitched cell is always in its 270° twist state, it compensates for any wavelength dependent effects in the switched cell when the switched cell is in the off state. The result is a darker off state and therefore an increase in contrast. More recent modifications of the STN display are the *super birefringence twisted effect* (SBE) cell and the *optical mode interference effect* (OMI) display. Both of these utilize large twist angles.

Twisted nematic LCDs can be made colored by the addition of colored filters or colored polarizers. If a single color is used, the display appears brightly colored on a black background or vice versa. If two differently colored polarizers are placed at right angles in front of the cell with a normal polarizer behind the cell, the characters will appear in one color and the background in another color. The reason for this is that certain colored polarizers only effectively polarize light of one color, while light of other colors is passed without being polarized. This results in light of two colors with polarization directions at right angles entering the cell. Both are rotated by the cell and only one is passed by the normal polarizer, depending on whether the cell is on or off. In either case, only one color of light emerges from the cell. Some of these colored twisted nematic LCDs are quite appealing.

An interesting possibility in designing twisted nematic LCDs concerns the fact that some liquid crystals prefer to align with the director parallel to the electric field if the frequency of the alternating voltage is low, but prefer to align with the director perpendicular to the field if the frequency of the applied voltage is high. The cell can therefore be switched to its twisted state by a high frequency voltage and switched to its untwisted state by a low frequency voltage. This method of operation is called *dual frequency addressing*. The advantage of this technique is that the switching times are reduced, but the use of high frequencies results in higher operating voltages and an increase in the power consumed.

Liquid Crystal Displays Using Dyes

A dye is a substance that absorbs light of a certain wavelength, causing the light reflected from or transmitted through the dye to appear colored. Some dye molecules absorb light of a certain wavelength better if the light is polarized along one axis of the molecule. Such dyes are called *dichroic dyes*, and these are the ones used in liquid crystal displays.

When dye molecules that are elongated in shape are dissolved in a liquid crystal, they tend to orient along the director of the liquid crystal. This effect is called the *guest-host interaction*, in that the dye molecules are the "guest" molecules being oriented by the liquid crystal "host" molecules. The dichroic properties of certain dye molecules can be utilized for display purposes by applying electric fields to the liquid crystal and causing reorientation of both the liquid crystal and the dye molecules.

Consider the display cell shown in figure 6.7. Light passing through a single polarizer on the front of the cell polarizes the light along the director of a liquid crystal that contains some dissolved dye aligned homogeneously. If the dye absorbs better when the light is polarized along the long axis of the dye molecules, the off state shown in figure 6.7(a) appears brightly colored in transmission or reflection. If the liquid crystal prefers to orient along the electric field direction, the on state shown in figure 6.7(b) shows much less color in transmission or reflection, since the light is polarized perpendicular to the long axis of most of the dye molecules. The result is colorless characters on a colored background. Just the opposite effect occurs if the liquid crystal is aligned homeotropically and the liquid crystal molecules prefer to orient perpendicular to the electric field. This results in colored characters on a colorless background.

The advantage of the guest-host LCD stems from the fact that only one polarizer is present. This increases the brightness of the display and reduces the change in contrast as the viewing angle is varied.

A slightly different version of the guest-host LCD operates with no polarizers. A chiral nematic liquid crystal with a fairly short pitch acts as the host material and a dichroic dye is dissolved in it. The surfaces are treated so that the sample is composed of many domains, each with a helical axis pointing in some direction (see figure 6.8[a]).

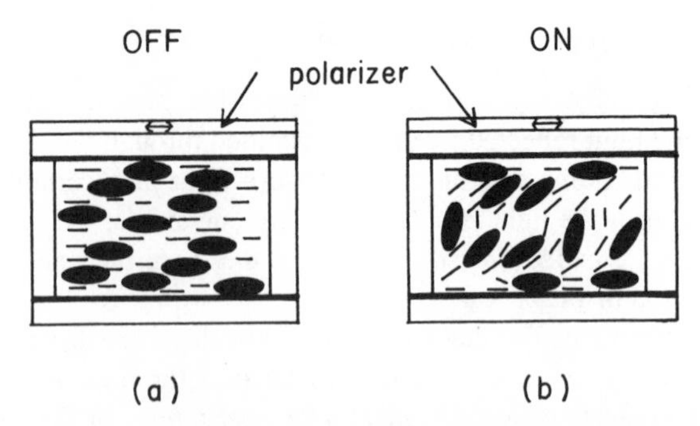

Fig. 6.7 Guest-host liquid crystal display using a dichroic dye. An electric field (b) causes a reorientation of the liquid crystal, which in turn reorients the dye molecules. The absorption characteristics of the cell therefore change from the condition with no applied electric field (a).

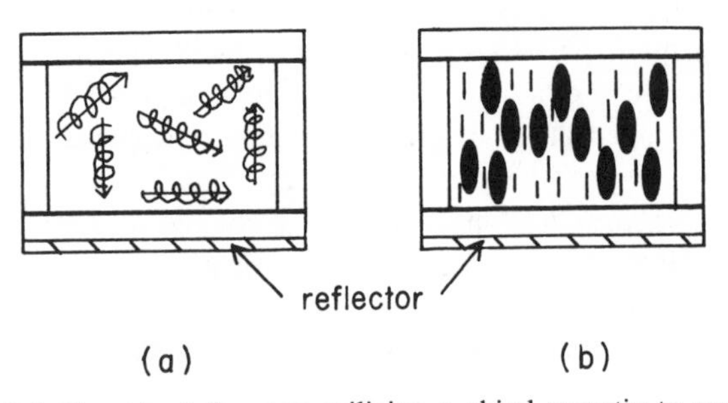

Fig. 6.8 Guest-host dye LCD utilizing a chiral nematic to nematic phase transition due to an applied voltage. The absorption characteristics of the dichroic dye molecules are different for the two phases.

The shortness of the pitch means that the director in each domain rotates around many times. This state is highly colored, because no matter what the polarization of the incoming light is, some of the dye molecules are oriented with their long axes parallel to the polarization direction. Assuming the liquid crystal molecules prefer to orient themselves parallel to an electric field, application of a voltage causes the dye molecules to lie perpendicular to the polarization direction for the incoming light. As evident from figure 6.8(b), this is actually a voltage induced chiral nematic to nematic phase transition. The result is a colorless on state against a brightly colored off state. Again, this type of display can be used in transmission or reflection. The lack of any polarizers makes this a very bright display, with superior characteristics over a wide range of viewing angles. In addition, the cell operates with a faster switching time in going from the on to the off state, since the twisting nature of the off state is due to the fact that a chiral nematic material is used and not simply due to surface alignment effects.

Birefringent Liquid Crystal Displays

In my discussion of birefringence, I explained how the phase angle between light polarized along one direction and light polarized along a perpendicular direction changes continuously as the light propagates through a liquid crystal. This effect can also be used to make a liquid crystal display.

Consider the situation shown in figure 6.9. The two polarizers are oriented at an angle of 45° to the director of the liquid crystal in the off state. As before, we therefore can think of the light coming out of the first polarizer as having two components: one polarized along

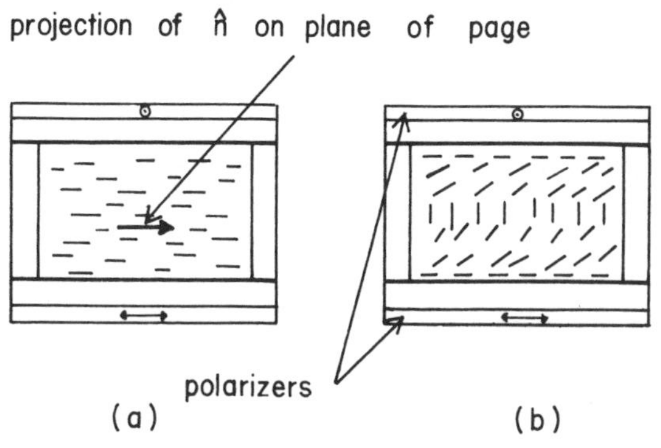

Fig. 6.9 A selective interference liquid crystal display. The polarizers are crossed and the director in the off state makes an angle of 45° with their axes. The phase angle change caused by the liquid crystal in the off state (a) allows light to pass through the second polarizer. The on state (b) produces little phase angle change, so little light passes through the second polarizer.

the director of the liquid crystal and the other polarized perpendicular to the director. As the light enters the liquid crystal, these two polarizations have a phase angle between them of zero. As they propagate through the cell, the phase angle increases, so at times the light is circularly polarized, linearly polarized, or elliptically polarized. If the thickness of the cell is just right, the light coming out of the cell is linearly polarized along the axis of the second polarizer (which is shown oriented perpendicular to the first polarizer). The light passes through and the display appears bright. If the liquid crystal material tends to line up with an electric field, the on state (figure 6.9[b]) shows no birefringence since the light polarized in any direction perpendicular to the director travels at the same velocity. There is no phase change so no light passes through the second polarizer. The result would be dark characters on a bright background, whether the cell is used in transmission or reflection.

Such cells are called *selective interference liquid crystal displays* because the phase angle of the emerging light strongly depends on wavelength. The off state is therefore brightly colored, since the phase angle is correct for only one wavelength. Unfortunately, this color depends on the temperature, viewing angle, and thickness of the cell. Two advantages of this display are that the voltage threshold is ''sharper'' and the switching times are somewhat faster than the twisted nematic LCD.

A slight variation of this display utilizes a liquid crystal that tends to orient perpendicular to an electric field. In this case the cell is always birefringent, but an applied voltage changes the phase angle of the emerging light by causing the order parameter of the liquid crystal to vary slightly. This changing phase angle produces variation in the color of light that is allowed to pass through the second polarizer. Large changes in voltage produce two distinct colors (an off and on state), but continuously variable colors are also possible for intermediate values of the applied voltage. This *tunable birefringence liquid crystal* display is therefore capable of producing beautiful and continuously variable colors, which make it perfect for some projection-type displays.

To avoid the problems of color uniformity due to variations in the thickness of the cell, viewing angle changes, and temperature drifts, the birefringent effect can be combined with a twisted nematic LCD.

A sheet of solid birefringent material of uniform thickness (calcite, for example) is placed between the liquid crystal and the polarizer of a twisted nematic cell (see figure 6.10). Light from above, after traversing the twisted nematic cell, is polarized either parallel or perpendicular to the initial polarization direction depending on whether the cell is off or on. In either case, this direction is at 45° to the axes of the solid retardation sheet. The retardation plate introduces a phase shift of 180°, which produces light polarized at an angle of 90° relative to the light that entered the retardation plate. This light either passes through the second polarizer or does not, depending on whether the twisted nematic cell rotated the light from the first polarizer. Actually, the retardation plate works this way for only one wavelength (color) of light, so one sees one color in the off state and the complementary color in the on state. This display is therefore a two color display, with the two colors being complementary. Since the birefringent material is a solid, its thickness is quite uniform and the phase angle change does not depend strongly on temperature. The viewing angle problem remains, and is in fact slightly worse than in twisted nematic LCDs.

Smectic Liquid Crystal Displays

Smectic liquid crystals usually do not respond to applied electric and magnetic fields as easily as nematic liquid crystals do. For this

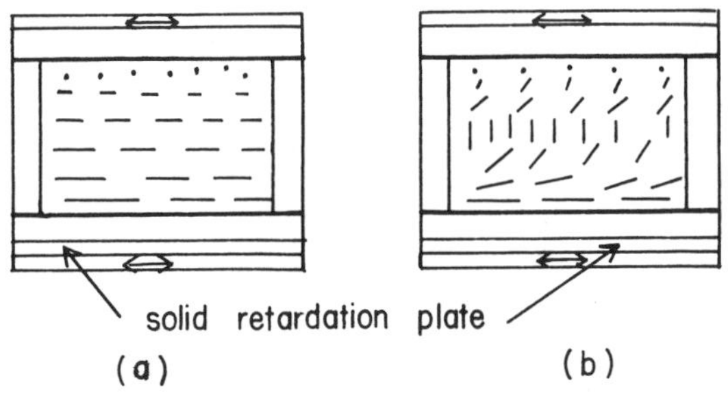

Fig. 6.10 LCD with a solid birefringent plate. The twisted nematic LCD either rotates the plane of polarization by 90° (a) or does not (b). The solid retardation plate causes a phase angle change, which either does or does not allow the light to pass through the second polarizer.

reason, simple application of an electric field rarely produces enough change in the optical properties of a smectic liquid crystal to be useful for display purposes. However, the texture of a smectic liquid crystal does depend strongly on whether it cooled to form the smectic phase in the presence or absence of an electric field. For example, a liquid crystal that prefers to align its director along an electric field will cool into a smectic phase with the layers parallel to the glass surfaces if the electric field is present during cooling. This is shown in figure 6.11(a). This texture is very uniform and scatters little light. If no electric field is present during cooling, however, the smectic liquid crystal texture is unaligned, with the smectic planes oriented in different directions in different parts of the sample. This is called the *focal conic texture* and is shown in figure 6.11(b). This texture strongly scatters light, so the cell appears milky white. If dye molecules are present in the smectic liquid crystal, colors other than white can be produced. Both the clear and scattering textures remain after the cooling process, even if the electric field is removed. This means that smectic LCDs do not need to be frequently refreshed, so they are ideal for high degrees of multiplexing. In addition, no polarizers are necessary, so these displays have good contrast and can be viewed at wide angles.

Smectic LCDs have been addressed in two ways. In the first method, a voltage that causes local heating to the higher temperature liquid crystal phase is applied to a row. Immediately following this heating pulse, voltages are applied to all the columns that should be clear in that row. Pixels with applied voltages cool in the presence of

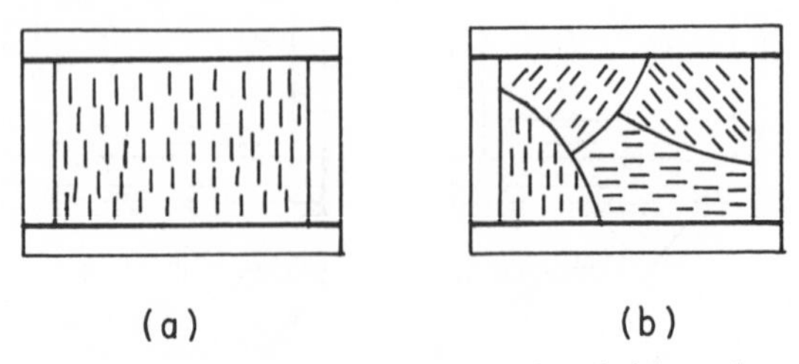

Fig. 6.11 Smectic liquid crystal display. If cooled from the nematic phase in the presence of an electric field, the state shown in (a) forms and is transparent. Cooling without an electric field produces the state shown in (b), which is highly scattering.

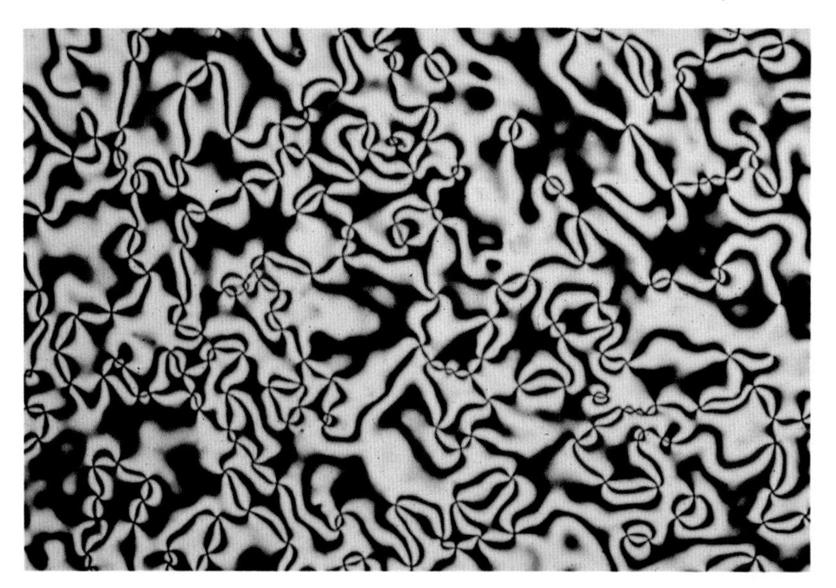

Plate 1. Nematic liquid crystal. The liquid crystal sample is between crossed polarizers. The liquid crystal appears dark where the director points along one of the polarizer axes. Points where dark areas converge are disclinations. (Courtesy of E. Merck Company)

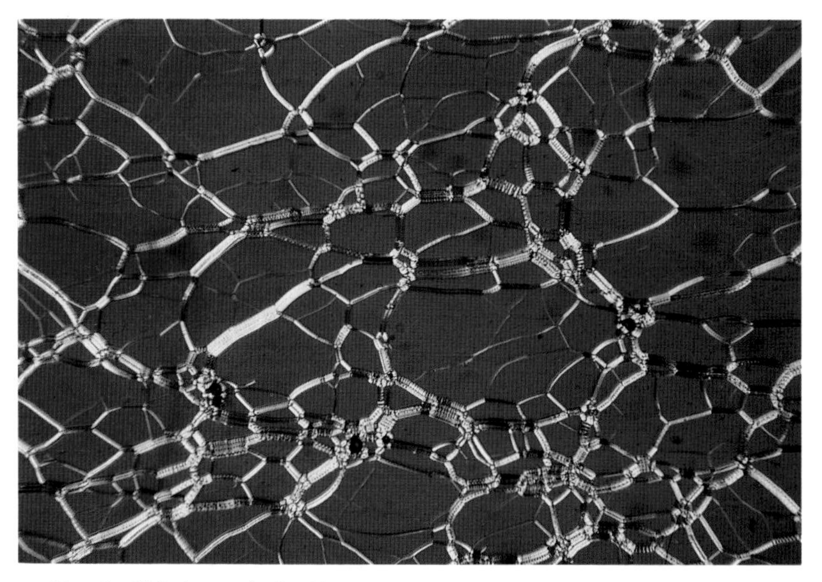

Plate 2. Chiral nematic liquid crystal (Grandjean texture). The liquid crystal sample is between crossed polarizers and the helical axis is perpendicular to the picture. The white lines running through the blue background are disclinations. (Courtesy of H. Kitzerow)

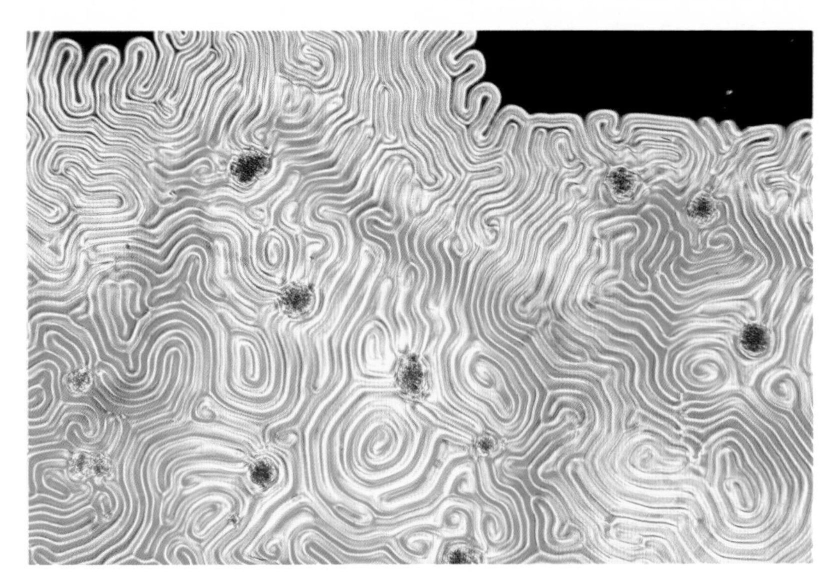

Plate 3. Chiral nematic liquid crystal (Fingerprint texture). The sample is between crossed polarizers and the helical axis is in the plane of the picture. The lines are individual turns of the helix. (Courtesy of H. Kitzerow and G. Heppke)

Plate 4. Smectic *A* liquid crystal. The liquid crystal sample is between crossed polarizers. (Courtesy of M. Neubert)

Plate 5. Smectic *B* liquid crystal. The liquid crystal sample is between crossed polarizers. (Courtesy of J. Goodby)

Plate 6. Discotic liquid crystal. The liquid crystal sample is between crossed polarizers. The liquid crystal phase (bright area) is forming from the liquid phase (dark area). (Courtesy of J. Goodby)

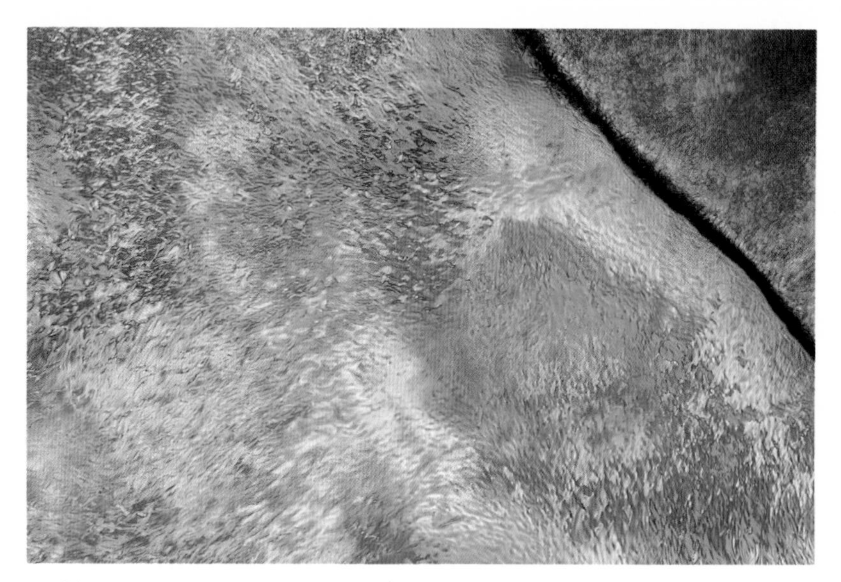

Plate 7. Polymer liquid crystal. This polyester sample is between crossed polarizers and in the nematic phase. (Courtesy of R. Cai, H. Toriumi, and E. Samulski)

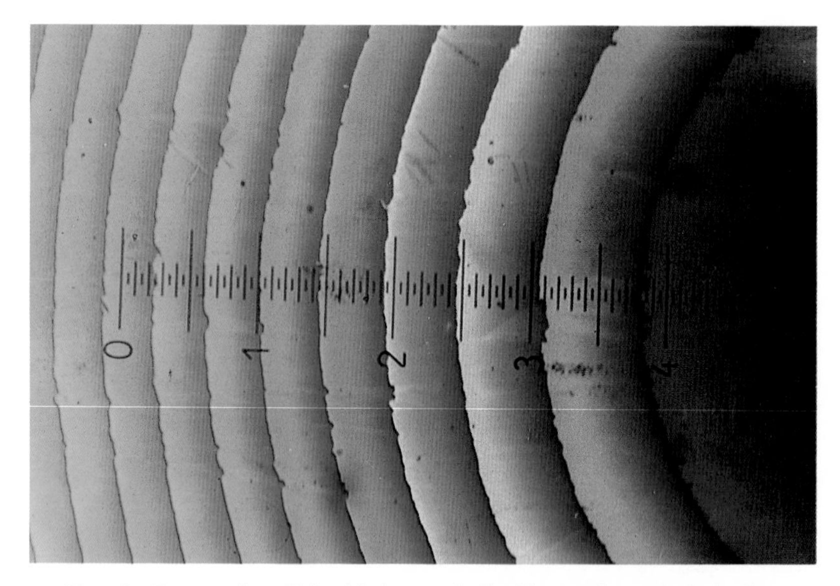

Plate 8. Cano wedge. This chiral nematic liquid crystal sample is in the space between a flat piece of glass and a curved lens, which is in turn between crossed polarizers. The circular lines are disclinations. (Courtesy of A. Feldman and P. Crooker)

Plate 9. Switchable light panels. The middle panel is in the opaque state, while the two outside panels are in the clear state. (Courtesy of Taliq Corporation)

Plate 10. Laptop computer with liquid crystal display. The display is one of the more recent supertwist variety and is lit from the rear. (Courtesy of Tandy Corporation/Radio Shack)

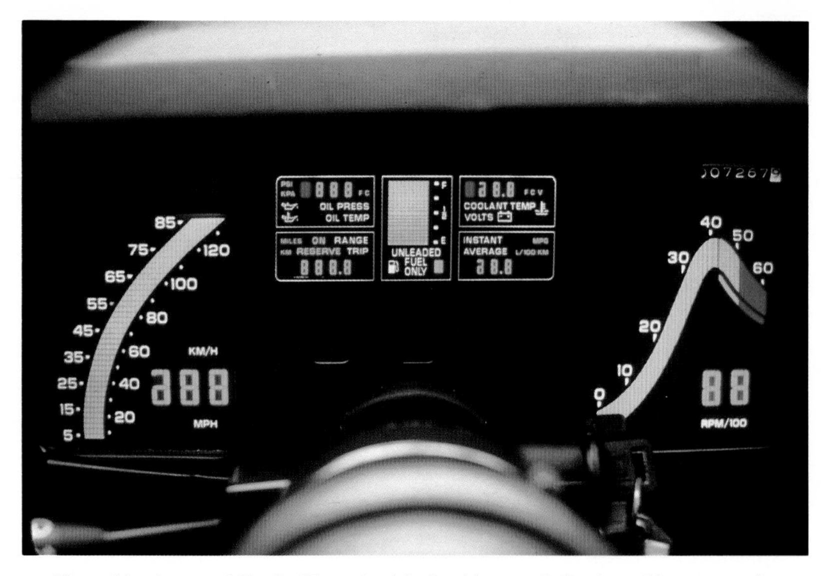

Plate 11. Automobile dashboard with liquid crystal display. (Courtesy of General Motors Corporation)

Plate 12. Pocket color television with liquid crystal display. This display uses the active matrix technology. (Courtesy of Citizen Watch)

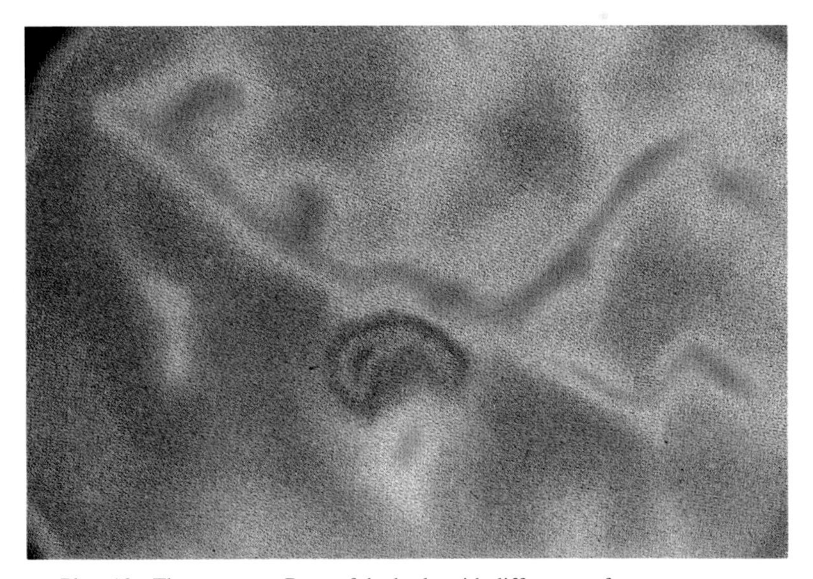

Plate 13. Thermogram. Parts of the body with different surface temperatures show up as different colors. (Courtesy of Qmax Technology Group)

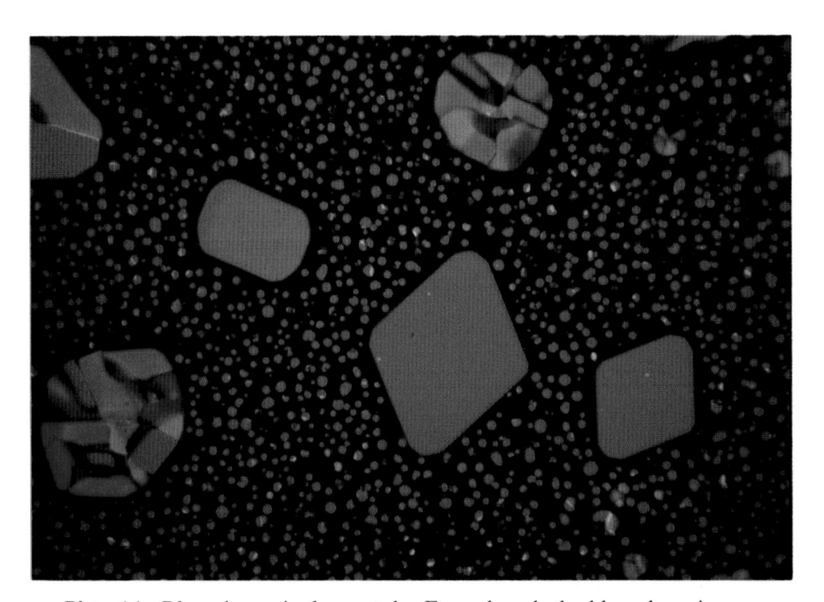

Plate 14. Blue phase single crystals. Even though the blue phase is completely fluid, its red and green crystals sometimes grow with flat boundaries. In some cases facets are visible. (Courtesy of H. Stegemeyer and Th. Blumel)

Plate 15. Cross-hatched blue phase single crystal. The cross-hatching appears after a phase transition from one blue phase to another. (Courtesy of H. Stegemeyer and H. Onusseit)

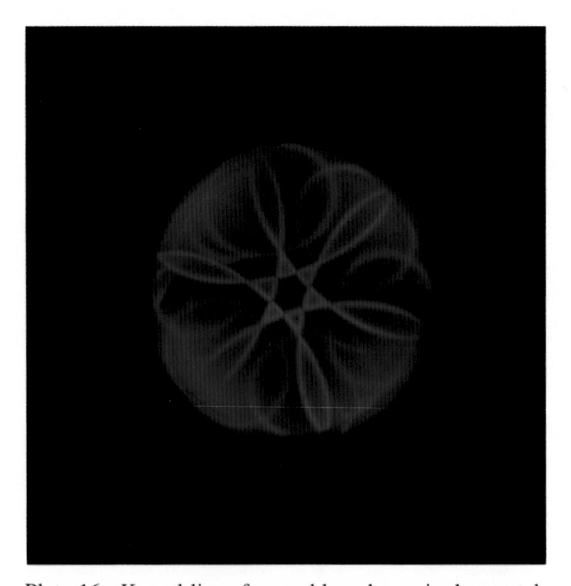

Plate 16. Kossel lines from a blue phase single crystal. These lines appear under certain viewing conditions and clearly reveal the symmetry of the single crystal. (Courtesy of H. Kitzerow and P. Pieranski)

an electric field and appear transparent while the other pixels of that row cool without an applied electric field and appear colored. The heating pulse is then applied to the next row, followed by the application of voltages to the proper columns. This process continues for all rows of the display and then repeats itself. The second method employs a laser. First, the entire display is quickly heated and then allowed to cool in the presence of an electric field. Second, a laser beam scans across the clear display, heating the liquid crystal wherever it hits. Since the electric field has been removed, the areas heated by the laser beam cool with no electric field applied, and thus appear colored. Although the information content of these laser addressed smectic LCDs can be extremely high, the laser itself and its projection optics are quite expensive. Still, these displays are useful for large area projection situations. Whether thermally or laser addressed, the need for electrical power to heat the liquid crystal eliminates one of the chief advantages of most LCDs—low power consumption.

Another type of LCD uses a chiral smectic *C* liquid crystal instead of a smectic *A*. As shown in figure 5.11, chiral smectic *C* liquid crystals possess a director that rotates in a cone in going from one smectic layer to the next. However, if the cell is thin enough, the interaction of the liquid crystal with the surfaces produces a texture in which there is no rotation of the director within the cell. If the director prefers to align parallel to the glass, the two states shown in figure 6.12 are possible. In both of the states shown in figure 6.12, the smectic layers are in the plane of the page and the director is parallel to the glass surfaces but pointing at an angle of 22.5° to the line perpendicular to the smectic planes. Even though the liquid crystal is not twisted, the fact that the phase prefers to be chiral means that the molecules preferentially orient one of the axes perpendicular to the long molecular axis. If the molecules possess a permanent electric dipole moment along this direction, then the liquid crystal itself possess a polarization perpendicular to the director. A phase with a permanent polarization in the absence of an electric field is called a *ferroelectric* phase, so this cell is a *surface stabilized ferroelectric liquid crystal device* (SSFLC).

To understand how this device can switch from one orientation to another, notice that each of the two states shown in figure 6.12 pos-

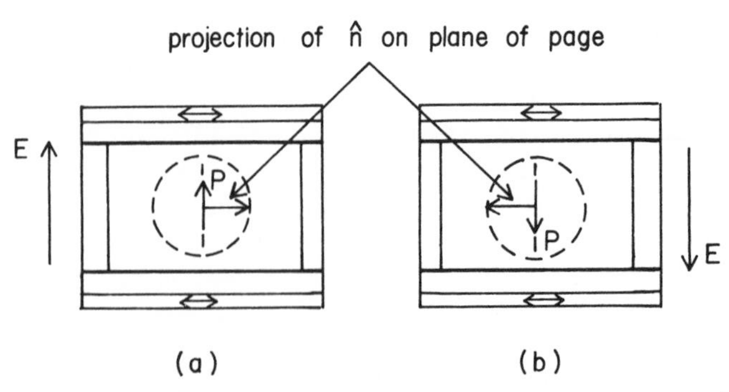

Fig. 6.12 Surface stabilized ferroelectric LCD. The smectic planes of the smectic C^* liquid crystal are in the plane of the page. The director is at an angle of 22.5° to the line perpendicular to these planes, and is switched between the two states shown by the action of an electric field on the permanent dipoles across the molecules. The polarizers are parallel and perpendicular to the director in (a) so this state is bright. The director is at an angle of 45° to the polarizer axes in (b), so the birefringence of the liquid crystal causes a 180° phase angle shift. This changes the polarization angle by 90°, causing the cell to become dark.

sess a polarization perpendicular to the surfaces. If an electric field is applied by transparent electrodes on the glass, the polarization responds by aligning with the field; an electric field in one direction causes the liquid crystal to adopt the state shown in figure 6.12(a), while an electric field in the opposite direction produces the state shown in figure 6.12(b). If crossed polarizers are present and aligned parallel and perpendicular to the director in the state shown in figure 6.12(a), then this state appears dark since the incoming light is polarized along the director and no phase retardation occurs. However, if the tilt angle is 22.5°, the light entering the liquid crystal state shown in figure 6.12(b) is polarized at an angle of 45° to the director. The two components of the light (parallel and perpendicular to the director) therefore suffer a phase shift in passing through the cell. If the thickness of the cell is adjusted so a 180° phase shift results, the emerging light is linearly polarized, but at an angle of 90° to the incoming light. The cell therefore appears bright. The switching times for these displays are faster than for other LCDs by a factor of one hundred to one thousand, because (1) the sample possesses ordered permanent electric dipoles, and (2) an electric field forces both the on and off transitions. Thus ferroelectric LCDs may play a significant role in displays that must have very short response times. In addition, since no heating or laser beam is necessary, these displays use very little power.

Before the SSFLC display can be exploited commercially, orientation problems of the liquid crystal inside the cell must be solved. Some researchers are attempting to overcome this problem by allowing the ferroelectric liquid crystal to retain some of its twist. It is too early to tell if these *deformed helical ferroelectric* (DHF) displays are capable of improved performance.

Polymer Dispersed Liquid Crystal Displays

A new development in liquid crystal displays utilizes polymers to contain the liquid crystal material instead of flat pieces of glass. These displays are easily fabricated and offer some significant advantages over the more common type of LCD.

The liquid crystal material is contained in tiny droplets embedded in a solid polymer matrix as shown in figure 6.13(a). One might think this is difficult to achieve, but in fact it is quite easy. The key is to form the solid polymer by mixing a fluid polymer with a fluid cross-linking agent. The most common example of this is an epoxy that comes in two parts (resin and hardener). Once mixed, chemical reactions cause cross-links between the normally fluid polymer to form, resulting in a solid material. If a liquid crystal material is dissolved into the two-part mixture while it is still fluid, the cross-linking reactions force the liquid crystal out of the solid polymer matrix and into fairly spherical droplets of almost pure liquid crystals. The size of the droplets is fairly uniform and can be varied from 0.1 to 10 micrometers in diameter by adjusting the rate at which the cross-linking reaction takes place. If the polymer and liquid crystal harden between two parallel surfaces, a *polymer dispersed liquid crystal* (PDLC) film results.

These films normally scatter light extremely efficiently. The reason for this can be seen in figure 6.13(b). The liquid crystal director inside each droplet adopts one of two configurations, depending on whether the liquid crystal prefers to align parallel or perpendicular to

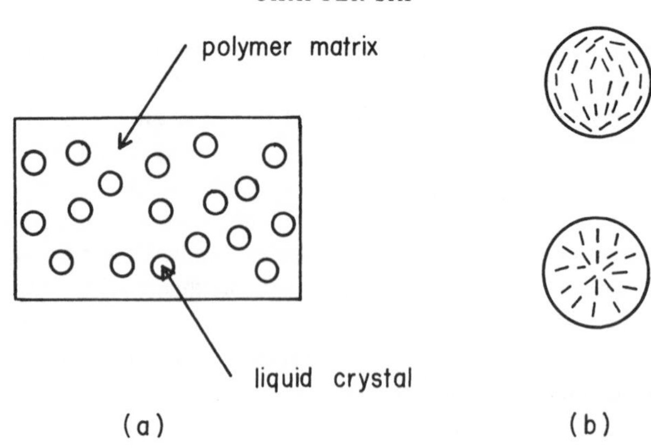

Fig. 6.13 Polymer dispersed liquid crystal display. The droplets of liquid crystal are shown dispersed throughout the polymer in (a). The two possible director configurations inside the droplets are shown in (b).

the polymer surface. In the case of parallel alignment, two points where the director is undefined (disclinations) occur, while the case of perpendicular alignment causes only one disclination to form in the center of the droplet. Let us concentrate on the case of parallel alignment. If no electric field is applied, each droplet is oriented with the line connecting the two disclinations oriented completely at random. Remember that light polarized parallel to the director travels at one velocity in the liquid crystal (index of refraction $n_\parallel$) and light polarized perpendicular to the director travels with a different velocity (index of refraction $n_\perp$).

The solid polymer is isotropic, so it has a single index of refraction (n_s) that in general is different from either of the liquid crystal indices. Light traveling through the PDLC film has its electric field oriented parallel to the surfaces of the film, so it must interact with droplets oriented with the director parallel to the electric field, perpendicular to the electric field, and every angle in between. It is impossible for the index of the polymer n_s to equal both $n_\parallel$ and $n_\perp$, so the sudden change of index causes reflections from most of the droplets. As evident from figure 6.14(a), a PDLC film with no electric field applied scatters light intensely.

However, if an electric field is applied across the PDLC film and the liquid crystal molecules prefer to align parallel to the field, the director configuration in each droplet is oriented so the line connecting the two disclinations lies parallel to the electric field. This is shown in figure 6.14(b). Light passing through the cell now has its electric field nearly perpendicular to the director, so if this index ($n_\perp$) is made equal to the polymer index (n_s), little reflection occurs at the boundary of each droplet. The PDLC film appears clear. To optimize the light scattering properties of these films, researchers carefully select the proper materials, droplet size, and film thickness.

There are two major advantages to these displays. First, no polarizers are necessary so these displays appear quite bright. Second, the thickness of the film is not critical, unlike with most other LCDs. Large-scale displays are easily fabricated and can even be flexible. One chief disadvantage is that the contrast declines for viewing angles away from perpendicular, because light traveling in these directions does not have its electric field perpendicular to the director (for which the index of the polymer is matched). Another disadvantage is that the voltage threshold curve is not sharp, so a high degree of multiplexing is difficult with these displays. Perhaps the ease of large-scale fabrication will be the aspect that brings these PDLC displays to the market. Switchable windows in buildings and automobiles, switchable partitions in office spaces, and large advertising signs

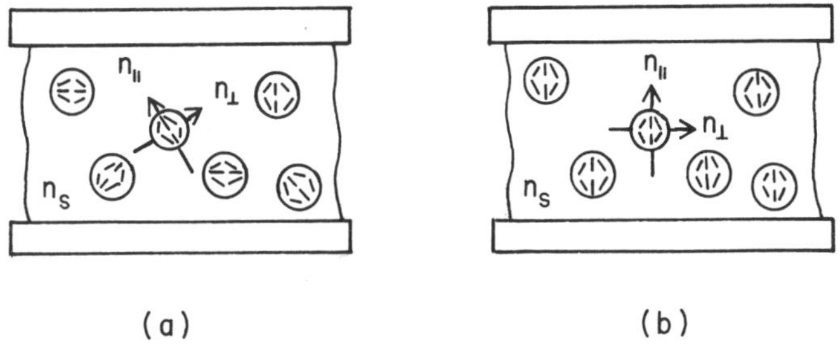

Fig. 6.14 Reorientation of the director within the liquid crystal droplets due to an electric field. The director inside the droplets is unaligned with no electric field present (a) so mismatch between the droplet index of refraction (namely $n_\parallel$) and the polymer index of refraction n_s scatters light. The director inside the droplets is aligned by the electric field (b) so equal indices of refraction ($n_\perp$ and n_s) reduce the scattering of light and produce a clear state.

(constantly changing!) are not far from becoming a reality. Plate 9 is a picture of a switchable window that is currently available. This window is actually manufactured in a two-step process call *microencapsulation*. First, tiny droplets of liquid crystal are surrounded with a thin coating of polymer. These encapsulated droplets are then introduced into a plastic or polymer matrix. Such films have properties very similar to the PDLC films.

As evident from the discussion in this chapter, the field of liquid crystal displays is a technology with many possibilities. In the last ten years, for example, the market for LCDs has increased about 25% each year, reaching over one billion dollars in 1988. One interesting aspect of this exponential growth is that LCDs are not necessarily replacing other displays that work well in some applications; rather, LCDs are being used in new ways and therefore opening new markets. At present, the LCD is second in sales only to the cathode-ray tube, and efforts to improve the quality of LCDs to the point where they rival the CRT are intense at research facilities all over the world. When that time comes, flat panel LCDs will replace the ubiquitous cathode-ray tube and the market for liquid crystal displays will surely explode. A television will no longer take up space in the corner of a room. Instead, a flat panel will hang from the wall, be propped up against a bookshelf, or simply stand on a table like a picture frame! Plates 10, 11, and 12 show products that utilize the technology described in this chapter.

CHAPTER 7

LCD Technology and Other Applications

The possibility of using liquid crystals in various types of displays was successfully demonstrated almost thirty years ago, yet it took ten to twenty years to develop the technology to manufacture LCDs of high quality and reliability. The first part of this chapter discusses the progress that has been made in the field of liquid crystal displays, ending with a description of recent efforts to utilize semiconductor technology in LCDs. Although displays are the most important application for liquid crystals at this time, other possibilities exist and some of these have been fully developed. These applications are discussed in the latter part of the chapter, and include descriptions of devices such as light valves and temperature sensors.

MATERIALS

As discussed in chapter 1, molecules that form liquid crystalline phases are elongated and fairly rigid in the central portion of the molecule. This is usually achieved by linking two or more benzene rings together in a linear arrangement. Flexible hydrocarbon chains attached to each end of the central benzene rings are also necessary to form liquid crystalline phases. The linkage group between the benzene rings is probably the most important part of the structure, so scientists usually refer to it by name. Hydrocarbon chains of various lengths can be added to the central portion, resulting in a series of liquid crystalline compounds. These different compounds possess different liquid crystal phases, giving the scientist or engineer many choices of materials for study or use.

One example of a liquid crystal series is based on one of the earliest liquid crystal molecules of significant importance. Figure 7.1 shows the first three members of the *p*-azoxyanisole (PAA) series. The central linkage between the two benzene rings is the azoxy- group for all members of the series. Only the number of carbon and hydrogen atoms in the hydrocarbon chains on each end of the molecule vary from one member of the series to the next. Both the temperature at which the liquid crystal phases form and the type of liquid crystal phase that forms vary from one member of the series to the next. This is typical of a liquid crystal series.

One way to illustrate this is to make a graph showing the transition temperatures as the number of carbon atoms in the chains is increased by one. Such a graph for the PAA series is shown in figure 7.2. Two features of this graph are common to many liquid crystal series. First, the temperature at which the liquid crystal phases are stable slowly decreases as the length of the hydrocarbon chains is increased. Second, the smectic phase is more likely to be present if the hydrocarbon chains are longer. Obviously, the presence of long hydrocarbon chains produces interactions between the molecules that promote the formation of layered structures.

In the PAA series, an oxygen atom connects the hydrocarbon chain with the benzene ring. An entire series of liquid crystal forming compounds with slightly different properties results when the hydrocarbon chains are connected directly to the benzene rings. This fact is illustrated in figure 7.3, where a graph of the transition temperatures for this series appears. Notice that the transition temperatures are much lower, and that a smectic *A* phase is present instead of a smectic *C* phase.

The central portion of the molecule is probably the most important feature in determining what liquid crystalline phases are present. The length of the hydrocarbon chains on either end and how the chains are attached to the central portion of the molecule can then be varied to achieve a wide range of transition temperatures and to some extent change the phases present. Progress in the development of materials suitable for LCDs has therefore come from research into new ways of linking the benzene rings together in the central portion of the molecule.

This point is nicely illustrated by the efforts around 1960 to develop a room temperature liquid crystal. The high transition temperatures of the azoxy- and azo-series make them unsuitable for display applications, so the search for a series with lower transition temperatures was of utmost importance in the early development of LCDs.

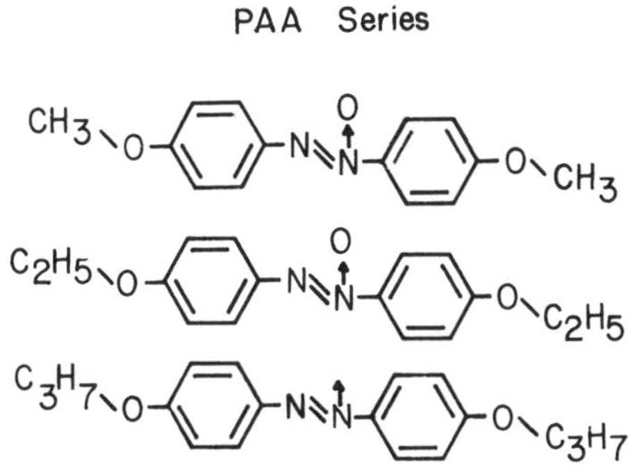

PAA Series

CH_3–O–C₆H₄–N=N(→O)–C₆H₄–O–CH_3

C_2H_5–O–C₆H₄–N=N(→O)–C₆H₄–O–C_2H_5

C_3H_7–O–C₆H₄–N=N(→O)–C₆H₄–O–C_3H_7

Fig. 7.1 First three members of the *p*-azoxyanisole (PAA) homologous series. The phases and transition temperatures of these compounds are given in the next figure.

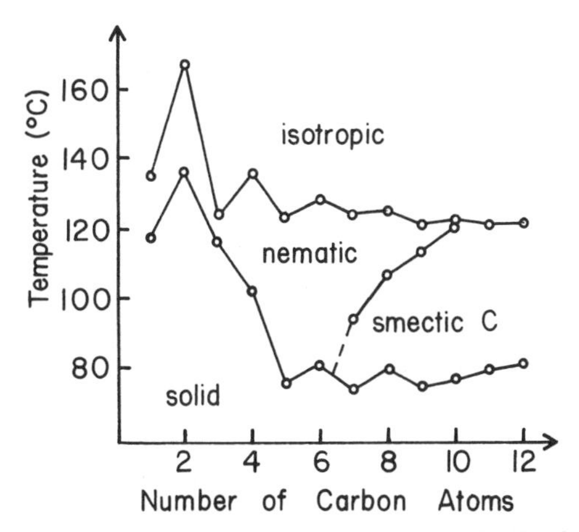

Fig. 7.2 The transition temperatures and phases for the first twelve members of the PAA homologous series.

The result (MBBA) has already been mentioned in chapter 2. The members of this series are called Schiff bases because of the central linkage group. Notice that MBBA has an oxygen atom linking one hydrocarbon chain to the benzene ring but no oxygen atom linking the other hydrocarbon chain. A series can therefore be formed by adding more carbon and hydrogen atoms to the oxygen linked chain only. The transition temperatures and phases for this series are shown in

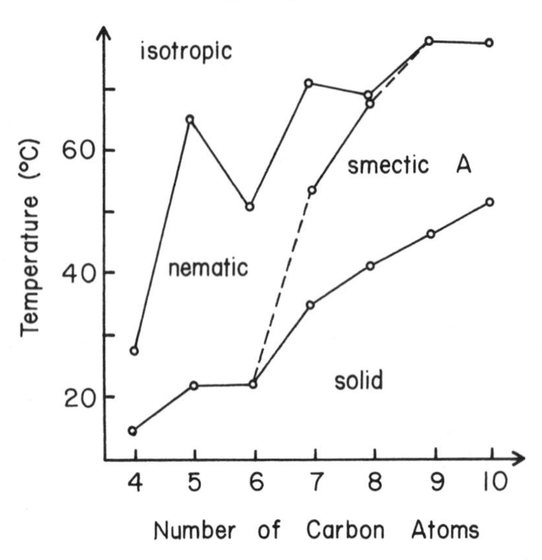

Fig. 7.3 The transition temperatures and phases for the fourth through tenth members of the alkylazoxybenzene homologous series.

figure 7.4. Note how low the transition temperatures are. Another series can be made by adding more carbon and hydrogen atoms only to the chain that is not connected to the benzene ring via an oxygen atom. Figure 7.5 shows the transition temperatures and phases for this series. The liquid crystalline phases occur at low temperatures in this series also.

One other central linkage group that was studied for use during the early development of LCD technology is the ester formed from benzoic acid. Two examples of these compounds are shown in figure 7.6. The transition temperatures of these compounds are fairly low, with the sequence of phases being different among members of the series.

Modern Materials

One of the most perplexing problems that haunted the development of materials suitable for LCDs was the lack of chemically stable liquid crystals. All of the compounds described so far break apart at the

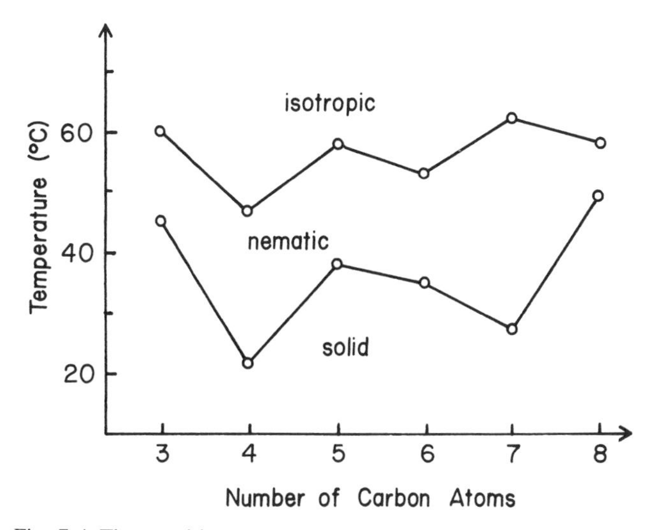

Fig. 7.4 The transition temperatures and phases for the third through eighth members of the homologous series created by changing the length of the oxygen containing chain only. The compound with four carbons is *p*-methoxybenzylidene-p-*n*-butylaniline (MBBA).

central linkage group when exposed to moisture, high temperature, and ultraviolet light. The pure compounds therefore become a mixture of the original compounds and new impurities. These impurities usually do not form liquid crystalline phases, so both the transition temperatures and the liquid crystal temperature ranges change as the impurities build up. Displays engineered to work effectively at room temperature thus begin to work less effectively, and can even cease to work if the liquid crystal phase is shifted above or below room temperature. Although liquid crystal displays were manufactured using these materials, they only worked for about a year or two.

The breakthrough was achieved in the early 1970s by a group of chemists working under George W. Gray at the University of Hull in England. They eliminated the central linkage group completely by connecting the benzene rings together, forming biphenyl and terphenyl compounds. These compounds are very stable, thus eliminating one of the chief problems hindering the development of high

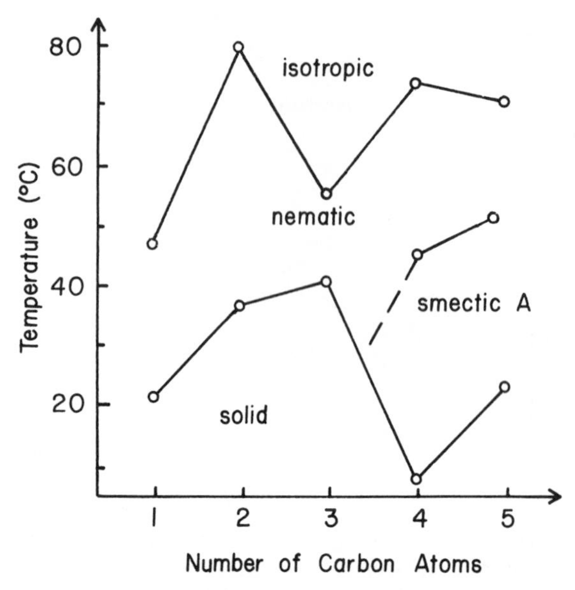

Fig. 7.5 The transition temperatures and phases for the first through fifth members of the homologous series created by changing the length of the chain that does not contain oxygen. The compound with one carbon atom is *p*-methoxybenzylidene-*p*-*n*-butylaniline (MBBA).

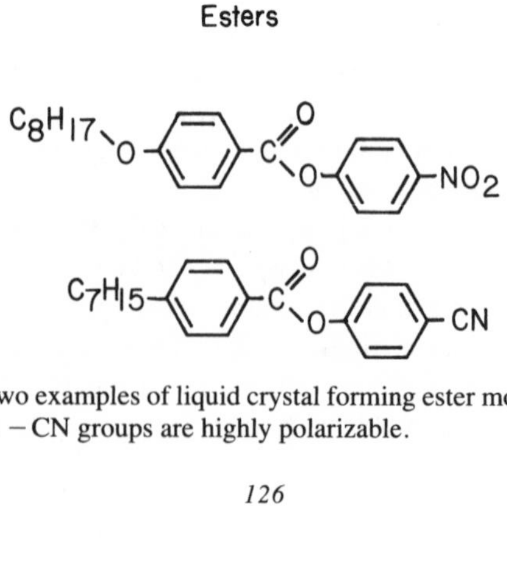

Esters

C_8H_{17}–O–C₆H₄–C(=O)–O–C₆H₄–NO_2

C_7H_{15}–C₆H₄–C(=O)–O–C₆H₄–CN

Fig. 7.6 Two examples of liquid crystal forming ester molecules. The $-NO_2$ and $-CN$ groups are highly polarizable.

quality LCDs. A few examples of compounds based on the biphenyl or terphenyl central core are shown in figure 7.7.

One important feature of some liquid crystals is illustrated by some of the compounds shown in figure 7.7. Notice that some contain a short polarizable group on one end instead of a hydrocarbon chain. In many cases the presence of such a group enhances the liquid crystalline character of the molecule, even though it makes the molecule shorter. The reason for this apparent paradox is that the highly polarizable group causes the molecules to pair up slightly, resulting in a combined molecule (called a *dimer*) that is longer than the single molecule. This formation of pairs was briefly discussed in chapter 5 where it was pointed out that X-ray studies showed that in some cases the smectic layers have a width equal to one and a half molecular lengths (see figure 5.10). The most common polarizable groups used in developing liquid crystalline materials are the cyano (−CN) and nitrous dioxide (−NO_2) groups. The transition temperatures for some of these liquid crystals are also quite low.

Recently research into replacing one or more of the benzene rings with cyclohexane rings has yielded promising results. These compounds are also very stable, resulting in the availability of entirely new series of liquid crystal compounds with different properties. Substitution within the benzene or cyclohexane ring is also an area of current research.

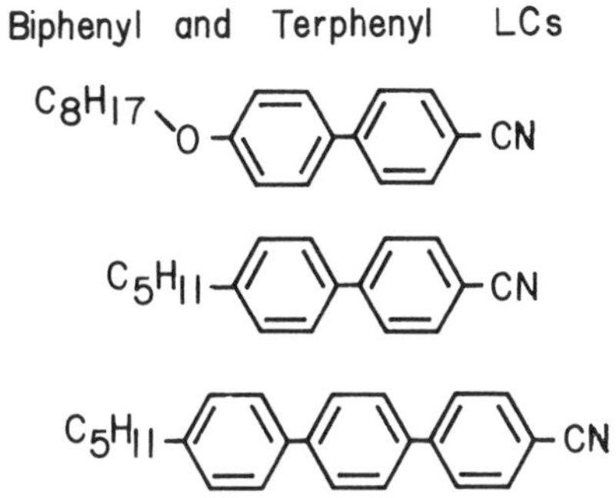

Biphenyl and Terphenyl LCs

C_8H_{17}–O–(biphenyl)–CN

C_5H_{11}–(biphenyl)–CN

C_5H_{11}–(terphenyl)–CN

Fig. 7.7 Three examples of biphenyl and terphenyl liquid crystal molecules, which are useful for display applications. The cyano (−CN) group is the polarizable end group.

CHAPTER SEVEN

SYNTHESIS OF LIQUID CRYSTALS

Liquid crystals are synthesized using the standard procedures of organic chemistry. By starting with available chemicals that resemble parts of the desired liquid crystal compound, chemical reactions are employed to link these parts together. For quite some time during the history of liquid crystal research, only very simple starting compounds were available. Many reactions were therefore necessary to obtain the final liquid crystal compound. More recently, complex starting compounds have become commercially available, so the synthesis proceeds with a fewer number of reactions.

Let us examine the synthesis of MBBA as an example of what actually must be done. MBBA can be formed in the reaction of *p-n*-methoxybenzaldehyde with *p-n*-butylaniline. Both of these compounds are shown in figure 7.8. The reaction produces MBBA (*p-n*-methoxybenzylidene-*p'-n*-butylaniline), which is also shown in figure 7.8, and water. This reaction is normally carried out in a mixture of solvents that provide the proper conditions for the reaction to take place, and the water produced during the reaction must be removed as the reaction proceeds. The MBBA is separated from the other compounds by filtering techniques and evaporation of the solvents. The final step is to purify the MBBA, usually by recrystallization in appropriate solvents.

In many cases the two compounds that react to form the liquid crystal compound are not commercially available. The chemist must first perform reactions that form these compounds, starting with other materials that are available. For example, *p-n*-methoxybenzaldehyde is produced in the reaction of *p*-hydroxybenzaldehyde (a compound similar to figure 7.8(a) but with an HO – group in place of the CH_3O – group) with *n*-methylbromide (CH_3Br). In this reaction hydrogen bromide is also given off. Likewise, *p-n*-butylaniline can be produced in a series of three reactions starting with *n*-butylbenzene (a compound similar to figure 7.8(b) but with simply a hydrogen atom in place of the $-NH_2$ group).

In general, these reactions are not simple ones to perform. If the conditions are not correct, the reaction does not proceed at all, or another reaction involving the starting material takes place, only to produce an undesired product. As if this is not enough, separation and purification of the product of these reactions are often not easy

(a) CH_3O–⬡–CHO

(b) H_2N–⬡–C_4H_9

(c) CH_3O–⬡–$CH{=}N$–⬡–C_4H_9

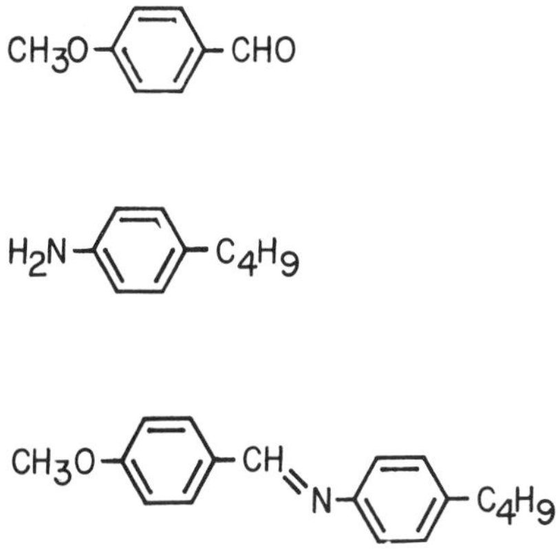

Fig. 7.8 Starting compounds and the final compound in the synthesis of MBBA: (a) *p-n*-methoxybenzaldehyde; (b) *p-n*-butylaniline; and (c) MBBA.

procedures. It takes an experienced organic chemist, skilled in these types of techniques, to efficiently synthesize liquid crystal compounds. Success is usually achieved after careful consideration of the details of the reaction together with some painstaking trial and error.

Mixtures

With so many liquid crystal forming compounds available, it might seem that finding one suitable for any application would be quite easy. This is simply not true. Any one application demands that the liquid crystal have the proper temperature range, the proper response to an electric field, the required stability, the proper viscosity, etc. For all the correct properties to occur in a single liquid crystal is improbable. Luckily, mixing two or more liquid crystals together allows the scientist or engineer to vary the properties of liquid crystals almost continuously. Many liquid crystals are miscible, with properties intermediate to the starting compounds. In many cases mixtures have wider temperature ranges for the liquid crystalline phase than either of the original liquid crystals. This is shown in figure 7.9 for a mixture of *p-n*-pentyl-*p*′-cyanobiphenyl (5CB) and *p-n*-octyloxy-*p*′-cyanobiphenyl (8OCB). 5CB has a nematic range of 24°–35°C while 8OCB has a nematic range of 67°–89°C, neither of

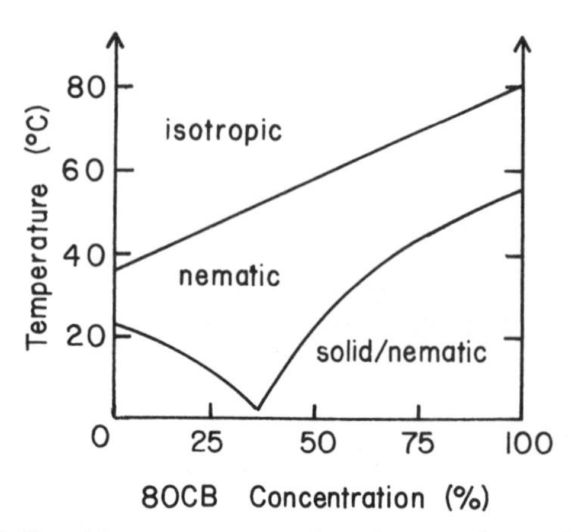

Fig. 7.9 Transition temperatures for mixtures of pentyl-cyanobiphenyl (5CB) and octyloxy-cyanobiphenyl (8OCB).

which is satisfactory for display purposes. However, a mixture of roughly 35% 5CB and 65% 8OCB has a nematic range of 5°–50°C, which is quite suitable for an LCD. A mixture for which the nematic range is as wide as possible due to a minimum in the solid-nematic transition temperature is called a *eutectic* mixture. The mixtures used in LCDs are eutectic mixtures, usually consisting of four to ten components. Two examples of fairly recent mixtures are E43, which is composed of biphenyls and terphenyls and has a nematic range from −10° to 84°C, and NP1694, which is composed of phenylcyclohexanes and biphenylcyclohexanes and has a nematic range from −20° to 86°C. The commercial availability of stable liquid crystal mixtures with superior electrical and optical properties plus a wide nematic range has probably been the most important factor in the recent development of high quality and reliable LCDs.

Dyes

As discussed in chapter 6, dichroic dyes are used in some types of liquid crystal displays. To be useful, a dye must be fairly soluble in the liquid crystal and not degrade the characteristics necessary for LCD application. In addition, the dye must be stable, uncharged, strongly absorb light, and produce colors suitable for displays. In order for the dye to have very different absorption characteristics in the off and on states, the dye molecules must be well oriented along the director when dissolved in the liquid crystal. The dye molecules themselves, therefore, must be elongated in shape and the liquid crystal itself must possess a high order parameter.

The two types of dyes that have found the most use in LCDs are those based on the azo- and anthraquinone groups. As mentioned before, the azo- group is chemically unstable so these dyes degrade over time. This fact is unfortunate, because otherwise the characteristics of the azo- dyes are superior to the anthraquinone dyes. Hopefully research going on right now will produce anthraquinone dyes or other dyes that are both stable and possess superior characteristics for display devices. Figure 7.10 shows a few of these azo- and anthraquinone dyes.

Other Factors in LCD Technology

In all LCDs, the liquid crystal is contained between two pieces of glass. Although the choice of the type of glass used is not critical,

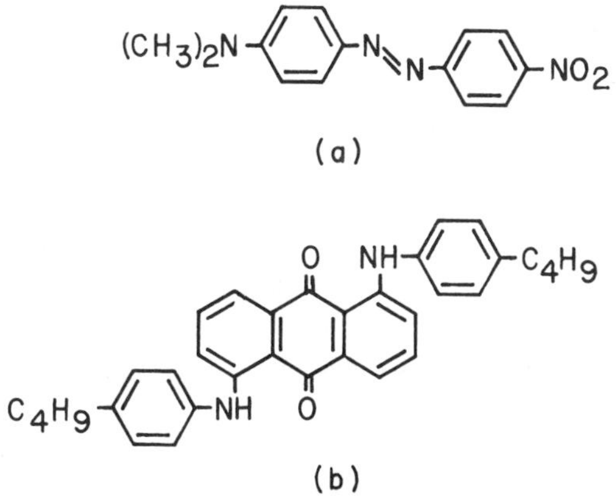

$(CH_3)_2N$ N N NO_2

(a)

O NH C_4H_9

C_4H_9 NH O

(b)

Fig. 7.10 Two examples of dichroic dye molecules useful for liquid crystal display applications; (a) azo- and (b) anthraquinone.

research over the years has shown that using glass that contains a significant amount of sodium or other alkali ions causes some problems. These ions migrate to the surface of the glass and together with any moisture present form a charged layer that can change both the alignment and electric field pattern within the cell. Two techniques have been used to avoid this problem. One is to use borosilicate glass, which contains few alkali ions. The other is to coat the glass with a layer of silicon dioxide that blocks the alkali ions from coming into contact with any moisture. This problem can also be overcome by using plastic instead of glass. This is an important area of current research, since plastics offer the other advantages of lower breakability, high uniformity, and flexibility. The main problems to be overcome concern the fact that low-cost plastics are usually birefringent and some react chemically with the liquid crystal compounds.

A very thin layer (10–50 nm) of indium tin oxide is usually used to make the transparent electrode pattern. These thin films transmit over 80% of the light incident on them and possess an electrical resistance low enough for proper operation of the cell. The transparent electrode material must form a precise pattern on both the front and rear glass surfaces. The two patterns are made to overlap only in the proper places. In this way the small strips of film that allow voltages to be applied to each element do not have an electrode across from them and therefore do not show up during operation of the cell. These electrode patterns are usually made by starting with glass fully coated with an indium tin oxide film. A mask is then made through either a silk-screening or photolithography process and then applied to the fully coated glass. The unwanted indium tin oxide is then chemically etched away. A technique that gives finer definition is to use glass that has an additional film of material resistant to etching on top of the indium tin oxide. A mask is made again, but this time the mask determines what parts of the glass will be exposed to ultraviolet light. The resistive layer loses its resistance to etching after exposure to ultraviolet light, so the glass can be placed in contact with the etching solution without the mask in place. Actually, resistive films come in two varieties. In the second type the original film is resistive to etching only after exposure to ultraviolet light, so the mask must be the negative of the desired electrode pattern.

The most common alignment for LCDs is the homogeneous texture where the molecules are parallel to the glass surface. This alignment is usually achieved by applying a thin coating of some long chain polymer and then stroking the coating in a single direction with a soft material. This process may produce many small grooves in the polymer, all pointing in the same direction, or may simply stretch the polymer film in one direction. These small grooves or stretching force the director to lie parallel to the direction of rubbing. Polyvinyl alcohol, some silanes, and polyimides are examples of coatings that work well. The first is susceptible to moisture, and the second produces such a thin layer that reliability is sometimes poor. The third is presently a popular alignment agent. One other technique for creating homogeneous alignment is to evaporate silicon oxide onto the surface using a very oblique angle. This process does not produce the low tilt angles that can be achieved by using the rubbed films and is not as convenient in the manufacturing process, especially for large-scale displays. This process does achieve reliable results, so it is used in the manufacture of most digital watch displays. Homeotropic alignment (where the liquid crystal molecules are aligned perpendicular to the glass surfaces) is usually achieved by coating the material with an amphiphilic material. The polar end adheres to the glass while the nonpolar end (usually a hydrocarbon chain) points into the liquid crystal space. These oriented hydrocarbon chains at both glass surfaces serve to orient the liquid crystal material perpendicular to the glass.

The thickness of the LCD is a critical factor in the proper operation of the cell. Not only does the critical field depend on the thickness of the cell, but in certain types of cells an optical effect that is determined by the thickness of the cell is essential to cell operation. The thickness of the cell (usually 5–25 μm) must therefore be exactly controlled. The sealing material around the outside of the cell is usually used as the spacer. Typical materials are thermoplastics and thermosetting organic polymers. This technique sometimes does not achieve the required thickness uniformity, so an additional measure is to place glass fibers or beads of the proper diameter in the liquid crystal material. These dispersed spacers are invisible to the eye but insure that the sealant material will cure with the cell at the proper thickness.

The polarizers used in many of the types of liquid crystal displays cause some of the most serious problems. For example, as a polarizer is made more efficient (polarizes the light better) it also transmits less light. The present compromise between brightness and contrast utilizes polarizers that transmit 40% to 45% of the light (remember that 50% is all that an ideal polarizer can transmit). These polarizers also lose some efficiency over time and can be easily scratched. At present, the polarizers are made from stretched polyvinyl alcohol films containing iodine that are placed between layers of cellulose acetate. If a colored polarizer is desired, the polyvinyl alcohol contains a dye instead of iodine. The polarizer is attached to the glass surface with an acrylic adhesive and covered by a plastic layer for protection. A reflective polarizer is made by the addition of a metal foil reflector. It is an interesting fact that these polarizers are an expensive part (materials and assembly) in the manufacture of LCDs. Research into higher efficiency, lower cost, and longer-lived polarizers is therefore important to the continued improvement in LCD technology.

Active Matrix Displays

One of the most exciting of the recent developments in LCD technology is the incorporation of semiconductor devices in liquid crystal displays. The idea is quite simple. If the liquid crystal cell itself does not have the optimum characteristics, simply combine it with an electrical device so the combination does have the optimum characteristics. This would be easy if only one liquid crystal cell were involved, but LCDs contain many elements, each of which must be modified with the additional electrical device. This task is made easier, however, by the advanced semiconducting industry, which can fabricate large-scale integrated circuits made up of devices small enough to satisfy the requirements of an LCD.

This technique is best illustrated by considering a large LCD matrix display. As discussed in the last chapter, multiplexing simplifies making the connections to the individual pixels, but the threshold characteristics of the liquid crystal cell limit the number of pixels that can be multiplexed. A nonlinear device can be used to improve the threshold characteristics. A nonlinear device is simply any electronic device for which the output depends on the input in a nonlinear way. For example, if a nonlinear device is placed in series with the liquid crystal pixel as shown in figure 7.11, the pixel will not turn on until the voltage across the nonlinear device reaches its threshold. If the nonlinear device has a very sharp threshold characteristic, the liquid crystal cell plus nonlinear device will also have a sharp threshold curve. In addition, the pixel will turn on at a voltage equal to the sum of the threshold voltages for the nonlinear device and the liquid crystal cell. The increase in both the sharpness of the threshold characteristics and the threshold voltage itself increases the degree of multiplexing possible. Such a system is called *active matrix addressing* with two-terminal devices, since the nonlinear devices have only two terminals. Notice that each pixel still requires only two connections. As also shown in figure 7.11, the nonlinear device is fabricated on the glass with the electrode structure. Examples of nonlinear devices suitable for this purpose are back-to-back silicon diodes and metal-insulator-metal devices.

Even more interesting is the use of three-terminal nonlinear devices instead of two-terminal nonlinear devices. The best example of a three-terminal device suitable for this purpose is a thin film transistor. Figure 7.12 shows the connections to such a device and how one might be fabricated on the glass substrate. The operation of an active matrix addressed display with three-terminal devices is very different. The three-terminal device acts as a switch; a small voltage applied to its control terminal causes a larger voltage to be applied to

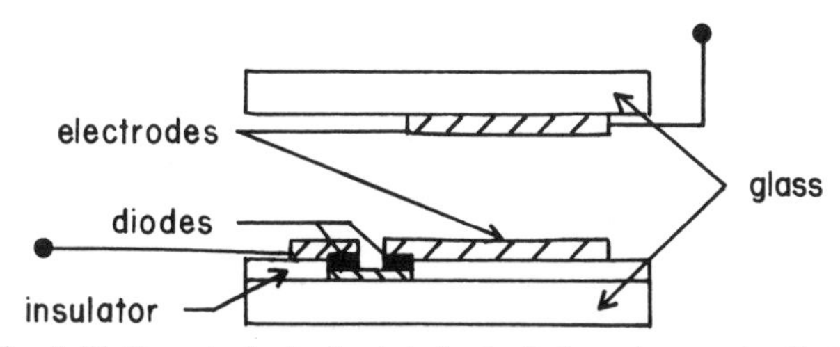

Fig. 7.11 One pixel of a back-to-back diode active matrix display. The liquid crystal is contained in the space between the two pieces of glass. The voltage is applied between the two electrodes shown with external connections and therefore across both the back-to-back diodes and liquid crystal cell in series.

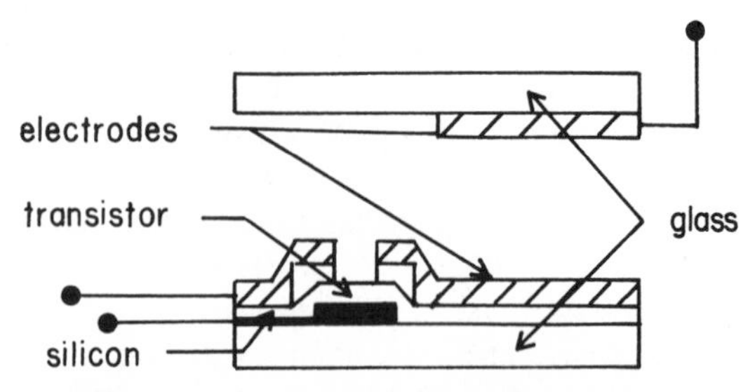

Fig. 7.12 One pixel of a thin film transistor active matrix display. The liquid crystal is contained in the space between the two pieces of glass. The external connections show the three elements of the display that are connected to other circuit elements.

the liquid crystal cell, thus turning it on. By using the proper three-terminal device, the liquid crystal pixel will stay on or off until the pixel is addressed again by the proper voltage to the control terminal of the three-terminal device. This effectively isolates the pixel from the voltages used to control other pixels. In addition, small variation of the voltage to the control terminal of the three-terminal device can cause precise variation of the brightness of the liquid crystal pixel. This type of display can therefore be engineered to produce an excellent gray scale.

The additional control available through the use of three-terminal devices obviously brings with it the additional complication of three connections to each pixel instead of two. Again, the highly developed technology of large-scale integration is capable of this much complexity. In fact, it is quite possible to drive these three-terminal active matrix displays directly, using a separate control line for each pixel. One idea that is currently being developed is to fabricate the electrical circuits necessary to send the proper voltages to these control lines on the glass substrate around the display itself. This would simplify making the connections between the display and other circuitry.

Liquid Crystal Light Valves

All liquid crystal displays are capable of varying the intensity of light emanating from different parts of the display. This feature has prompted researchers to investigate other possible applications besides displays. The general idea is that light of uniform intensity striking a liquid crystal panel could emerge from different parts of the panel with different intensities, if the director configuration of the liquid crystal varied from place to place within the panel. The name for any device that can be used to vary the proportion of the incident light emitted is a *light valve*. If the light valve has the capability of varying the proportion of emitted light from place to place, the device is then known as a *spatial light modulator* (SLM). Liquid crystal light valves (LCLVs) and liquid crystal spatial light modulators (LCSLMs) have many possible applications, including light amplifiers, large-scale projection systems, optical data processing, and visible-to-infrared light converters.

The area of optical data processing is important and worthy of some discussion. The recent explosion of computer technology has demonstrated the power of fast computing. While further enhancements in the speed of electronic computers are expected, there are inherent limitations on how fast electronic devices can operate. Already there are efforts underway to increase the speed of computers by having them perform operations in parallel (simultaneous operations utilizing separate devices). This technique also has its limitations. It is clear that optical computing with increases of speed by a factor of at least one thousand has the potential to revolutionize computer technology. Not only is the actual speed of a typical optical device greater than current electronic devices, the ease of creating parallel optical channels further confirms the potential of this new area of development. The heart of such a technology is the SLM, which optically performs the same functions as the electronic switching devices in present generation computers. The liquid crystal SLM is a promising candidate for such a system, and research into its use is now in progress.

Although most of the liquid crystal displays discussed in chapter 6 were electrically addressed (voltage signals controlled the display elements), we did see how a display could be optically addressed (scanning by a laser beam). In the case of liquid crystal light valves, however, much of the interest has been in optical addressing, since this method provides finer detail (higher resolution) and is faster. For example, consider the LCLV shown in figure 7.13. The liquid crystal cell is placed in series with a photoconductive layer. The electrical

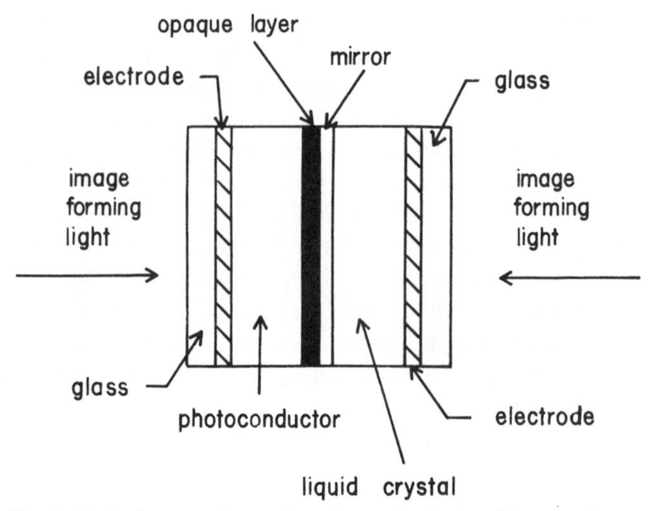

Fig. 7.13 A photoconductive image transfer device. The actual image is formed in the photoconductor by light coming from the left. The properties of the photoconductor cause an electric field pattern in the liquid crystal that is identical to the image. This image can then be seen when light strikes the liquid crystal from the right.

resistance of such a layer varies inversely with the amount of light striking it. If a voltage is placed across both the liquid crystal cell and the photoconductive layer, a small voltage is applied to the liquid crystal where little light strikes the layer and a large voltage is applied to the liquid crystal where a great deal of light strikes the layer. If the liquid crystal cell is designed to work in the reflecting mode, bright light incident from the other side will produce a bright image on this side that corresponds to the pattern of brighter light striking the other side. Notice that this is a *real time* image, since there is no delay necessary in producing the image. Any system that must scan across an image in order to process it is much slower.

Liquid crystal spatial light modulators need not be used just to capture an image. Similar devices can be used to process an image already captured. This is part of a field called *optical processing* and the advantage of LCSLMs is the speed at which they can perform various functions. As mentioned before, the liquid crystal devices in an optical processing system can be either electrically or optically addressed. For example, a liquid crystal light valve can be placed in series with an array of *charge-coupled devices* (CCDs) as shown in figure 7.14. Here a TV camera transfers the information it picks up in a scan to the CCD as voltage signals. The two-dimensional CCD array stores the information in the correct location as localized charge packets. Application of the proper control voltage to the CCD readout structure causes all of the charge to diffuse through to the liquid crystal cell, causing different amounts of deformation dependent upon the amount of charge present. In this way the entire two-dimensional image is transferred and read at once.

Liquid Crystal Displays of the Future

At this time the liquid crystal display market is second only to the cathode-ray tube (CRT) market, but the difference is enormous because CRTs are used in just about all televisions. Therefore, if LCD televisions attain the quality of CRT televisions, the LCD market will take off in a way difficult to imagine at this time. LCD color televi-

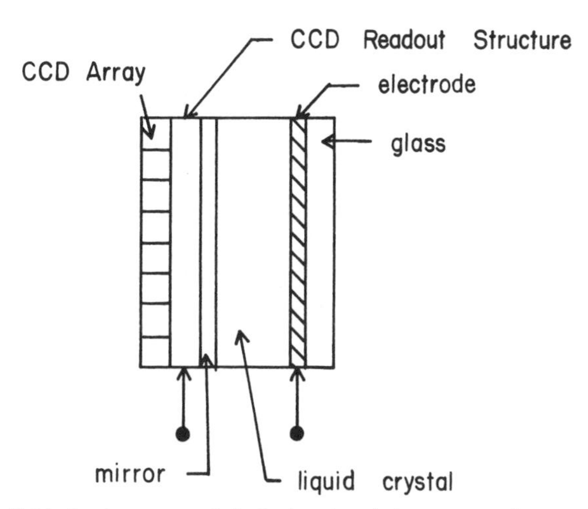

Fig. 7.14 A charge-coupled device (CCD) image transfer system. Light from the left forms an image in the CCD array that is transferred to the liquid crystal by the CCD readout structure. The liquid crystal image can then be seen when lit from the right.

sions using an active matrix display are already available in small, portable models. Plate 12 is a picture of such a product. The quality of the picture is quite acceptable, but not equal to that produced by a CRT. LCD televisions are now available with six-inch screens, but the size should increase to over ten inches within a few years. Among researchers in the field, twenty-inch flat panel televisions are expected to be on the market around 1995.

In the future, LCDs will be used more and more frequently in devices where the display is small or low power consumption is important. Watches, calculators, typewriters, telephones, and laptop computers will continue to use LCDs. We will begin to see them more often in automobile dashboards, portable instruments, and airplane cockpits. LCDs of large size will become available. For example, traffic signs and advertisement boards will offer the opportunity for large-scale signs to be easily and quickly changed. Tunable windows, which allow us to select the amount of light that passes through the glass may become common. Further advances will bring LCDs into the fast display or high information display markets. Optical information processing with liquid crystal devices is likely to become a new and substantial application. Likewise, large-screen projection systems using liquid crystal light valves could very well become a reality in the near future.

Liquid Crystal Temperature Sensors

One of the most interesting (and beautiful) ways liquid crystals can be used is to measure temperature utilizing the selective reflection property of chiral nematic liquid crystals. As discussed in chapter 5, chiral nematic liquid crystals reflect light with a wavelength in the liquid crystal equal to the pitch. If white light, which contains a continuum of wavelengths, strikes a chiral nematic liquid crystal and propagates parallel to the twist axis, only light with a wavelength in the liquid crystal equal to the pitch is reflected back. To an observer, therefore, the liquid crystal will appear to have the specific color as determined by the wavelength of light reflected. Since the pitch of chiral nematic liquid crystals changes with temperature, the color reflected by these materials also changes with temperature. In this way the temperature can be measured simply by observing the color of the chiral nematic liquid crystal. Such thermometers have been available for some time, and can be used to measure the temperature of a person's forehead, of the water in a fish tank, or of a room.

The usefulness of chiral nematic thermometers rests in the ease of designing devices that respond in the temperature range desired. Many chiral nematic compounds have a pitch that lies in the visible part of the spectrum, but the temperature at which this is true varies from one compound to the next. By mixing together the proper compounds, designers of liquid crystal thermometers can produce a material that reflects the different colors of the spectrum over whatever temperature range is desired.

In fact, very sensitive thermometers can be made in this way. Some chiral nematic liquid crystals possess a pitch that changes drastically over a very small temperature interval. Again, by mixing the proper liquid crystals, designers can produce a thermometer that reflects the visible spectrum over a temperature interval of less than one degree. Such thermometers are capable of revealing extremely small temperature differences and are useful in temperature mapping applications. For example, a film of such a material placed in good thermal contact with the skin reveals in vivid color the temperature profile of the skin. This can be an important diagnostic tool for medicine, since many local medical problems (for example, a tumor located slightly below the surface) possess a temperature different from the surrounding tissue. If such an abnormality is present, it will show up on the liquid crystal film as an area of different color. The ability to test for slight temperature differences and possibly avoid surgery is obviously very important. Plate 13 is a picture taken by a commercially available instrument.

Liquid crystal temperature sensors have been used as a biofeedback mechanism in psychological therapy situations. All that is necessary is that the sensor be placed in close contact with the skin (between two fingers for example). The advantage of the liquid crystal device is that it is much less expensive and easier to use than the more conventional monitors used for biofeedback (heart rate, blood pressure, EEG, etc.).

Such liquid crystal films can also be used for mapping the temperature profiles of almost any device. In some cases, such as precision ovens or electronic instrumentation, the information is important in order to find the proper design of the device. In other cases, the liquid crystal film can be used as a form of nondestructive testing. For example, poor electrical connections usually are slightly hotter than good connections. Although a device such as a printed circuit board may function properly, a poor electrical connection usually degrades over time to the point where the device no longer functions correctly. A chiral nematic film easily reveals a slightly warmer connection, thus allowing it to be fixed before the product is sold.

Some time ago, such liquid crystal materials were used in "mood rings." These rings reflected different colors depending on slight differences in temperature and were purported to indicate a person's different "moods." At least one manufacturer of children's toys has researched the idea of employing such liquid crystal sensors on figures or in games. The idea is that such figures could display "moods" or the strategy of a game might change during play depending on slight temperature changes. Some artists have also experimented with using chiral nematic liquid crystals as a medium. By using different mixtures, an artist can produce a work that displays different colors, but more important, the colors would change with slight variations in the temperature of the room!

An important part of this technology is the ability to produce films containing chiral nematic materials. Thin films allow for good thermal contact between the temperature sensing material and the object, and keep the liquid crystal from becoming impure through contact with other materials. These films are made through the process of microencapsulation, which was discussed in chapter 6. This technique gives the designer good control over the purity of the encapsulated material, which is crucial for the design of sensitive thermometers. Chiral nematic liquid crystals can be dispersed in a polymer film by mixing in the liquid crystal as the polymer sets (as is done in PDLC displays). Although a much easier technique, the purity of the encapsulated material cannot be controlled nearly as well, so this technique may have serious limitations for thermometry.

It is no mystery why many chiral nematic liquid crystals show drastic changes in pitch over such a small temperature interval. Such changes usually occur in the chiral nematic phase in a temperature interval just above the transition to a smectic phase and reveal some basic physics about phase transitions in general. The smectic phase possesses layers that must buckle in order to twist, so it is much more difficult for smectic liquid crystals to twist than for nematic liquids that do not possess layers. As the temperature decreases toward the chiral nematic to smectic phase transition, the molecules start to momentarily and only locally arrange themselves in layers. The temperature is too high for these layers to establish themselves permanently, but small groups of molecules begin to show some fleeting layer ordering. As the temperature gets closer to the smectic transition, the length of time this local layering persists gets longer and the size of the groups of molecules get larger. This makes it increasingly difficult for the chiral nematic liquid crystal to twist, so the pitch lengthens. Since all this can only happen near the smectic transition, the pitch change takes place over a tiny temperature interval just above the transition. At the transition, the entire material changes to a smectic liquid crystal, which usually has no twist at all (in other words, it has an infinite pitch). By mixing different materials together, the chiral nematic to smectic transition temperature can be made to occur at any temperature desired, so the temperature at which the chiral nematic reflects visible light is completely controllable. Proper selection of compounds in the mixture also allows the interval over which visible light is reflected to be varied in size. Figure 7.15 shows some data for a few mixtures prepared for commercial application. One type of thermometer uses dots containing different mixtures that respond over a temperature range of about two degrees, but at temperatures that differ by about two degrees. At any one temperature, therefore, only one dot shows any color. The temperature is shown by the number written next to the colored dot.

Other Applications

Some less important applications of liquid crystals rely on the general ability of liquid crystals to show their delicate response to changes in the environment. For example, chiral nematic liquid crystals change their pitch in response to pressure changes in much the same way they respond to temperature changes. Fairly large pressure changes are necessary to produce these pitch changes (many atmospheres), but a chiral nematic device kept at a constant temperature displays different colors for different pressures and therefore can be

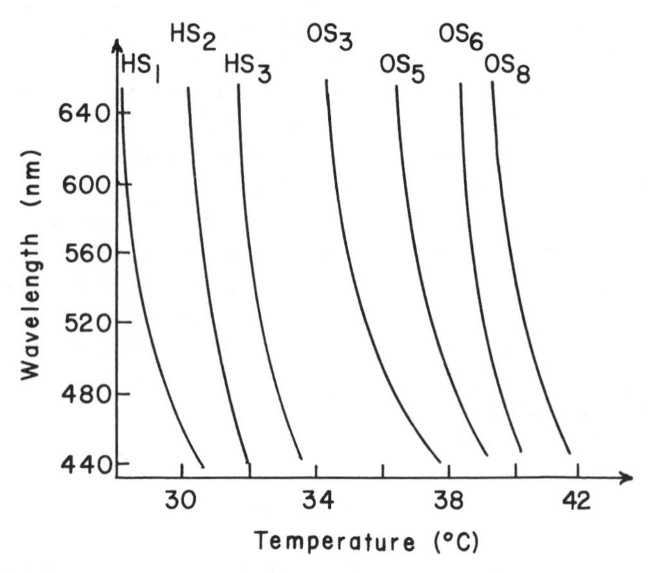

Fig. 7.15 Wavelength versus temperature curves for seven chiral nematic mixtures made for the purpose of application. Each mixture reflects all the colors of the visible spectrum over a two degree temperature interval, but each mixture does this at a slightly different temperature.

used as a pressure sensor. Likewise, exposure of chiral nematic liquid crystals to certain vapors causes a change in the pitch and a corresponding change in the color of light they reflect. Such devices can be used as sensors to detect low concentrations of these vapors. Ultrasonic waves striking an object cause its temperature to rise slightly. Therefore liquid crystal films can be used to detect the presence of these waves.

The Freedericksz transition discussed in chapter 3 is the basis for devices that can measure the strength of electric and magnetic fields. Since these fields affect the director configuration in a thin sample of liquid crystal, the properties of this thin sample are altered with a change in field strength. One can monitor an electrical, magnetic, or optical property of the sample (for example, its capacitance), and use it to measure the strength of the field. A device with a digital readout of the field strength is very feasible, although problems remain in making the device accurate over a long period of time.

Few liquid crystal applications utilize the mechanical properties of these phases, but at least one possibility is being explored. The viscosity of certain liquid crystal phases (for example, the smectic *B* phase) is much higher than the viscosity of the isotropic liquid phase. If such a material is used as the hydraulic element in brakes, clutches, or bearings, the amount of friction in these components depends on the phase of the material. In theory, temperature or pressure changes within the device could produce phase changes and thereby vary the characteristics of the device over quite a large range. The idea is to produce "smart" devices, which utilize the phase change to gain advantages not possible with conventional liquids. Our theoretical knowledge of exactly what happens when liquid crystals are used this way is lacking at this time, so a good deal of theoretical and experimental work will be required before these devices become a reality.

One last application of liquid crystals is a scientifically important one. In many instances, scientists and technicians perform measurements on substances that have been dissolved in some liquid. Two important examples come to mind. First, chemical reactions frequently are carried out in a liquid medium. Second, a sample containing various substances can often be divided up into its constituents by forcing the mixture to pass through a liquid. The rates at which the different components of the mixture move through the liquid are different, so the various components come out of the end of the apparatus at different times. This area of analytical chemistry is called *chromatography*, and it can be used to investigate what is in an unknown mixture or to actually separate a mixture into its component parts. This technique is extremely important to many of the analyses done in chemistry, biology, and medicine.

Since the solvents used in both of these situations are liquids, the molecules of the substance to be investigated are situated in an isotropic environment. Sometimes using an anisotropic solvent (i.e., a liquid crystal) changes how the substance behaves in a way that is extremely important. For example, chemical reactions sometimes occur differently in a liquid crystal than in an isotropic liquid. This allows chemists to control the reaction in a way not possible otherwise. Likewise, the rates different molecules pass through a liquid crystal are sometimes not the same, whereas through an isotropic liquid the rates are the same. Molecules that differ from each other in very minor ways can often be separated using a liquid crystal medium when other methods fail. Liquid crystals are being used with increasing frequency as an anisotropic medium for such studies, and the power of this technique is only now starting to be exploited.

Although these last few applications have been presented as less important, one cannot be sure this situation will always be true. The ability of liquid crystals to perform one of these functions may become the cornerstone for a future technology of huge proportions. Perhaps someone writing ten years from now about the applications of liquid crystals will devote more attention to one of these possibilities than to LCDs. Time will tell, but the delicate response of liquid crystals to almost any change in their environment ensures that researchers will find them useful for a wealth of purposes. How important they become and in what areas they become important are questions to be answered by scientists and engineers presently at work in laboratories all over the world.

CHAPTER 8

Lyotropic Liquid Crystals

In my discussion so far, I have concentrated on liquid crystal systems in which temperature is the variable that determines which phase of matter exists. Two other possibilities have been pointed out. The first was pressure, and throughout my discussion I have assumed that the external pressure exerted on the material is equal to atmospheric pressure. If the pressure on a sample is allowed to vary considerably, phase changes result just as if the temperature had been changed. In fact, pressure works in the opposite direction as temperature. Raising the pressure while keeping the temperature constant causes many of the same phase transitions one would observe if the temperature were lowered with the pressure constant. The second possibility I discussed was concentration in a mixture of two liquid crystals. As evident from figure 7.9, phase transitions result when the concentration of one component of a mixture is changed while the temperature is held constant. In all of these cases, however, temperature remains the most important variable in determining phase behavior—hence the name *thermotropic* liquid crystals for all of these materials.

There is another class of substances that displays liquid crystalline behavior only when mixed with another material. Although temperature is still an important variable in determining the phase present, the concentration of one component with respect to the other is far more important. These materials are called *lyotropic* liquid crystals and are just as interesting and important as thermotropic liquid crystals. For example, the ''goo'' that sometimes collects in the bottom of your soap dish is a lyotropic liquid crystal phase of the soap/water mixture. Likewise, every one of the cell membranes in your body owes its structure to the liquid crystalline nature of the phospholipid/water mixture. Lyotropic liquid crystals are both scientifically interesting and technologically important, as will become evident in the following discussion.

CHAPTER EIGHT

Amphiphilic Molecules

All of us have heard the statement that oil and water do not mix. Let us imagine ourselves as chemists, mixing various liquids together in hopes of gaining some insight into why certain liquids mix while others do not. The results of our investigation would show that oil and water are representative of two general classes of liquids. Liquids belonging to the same class usually mix, while those belonging to different classes do not. This statement is far too simplistic, because there are liquids that are not easily classified into one class, and other variables such as temperature turn out to be important. But as a general description of the mixing phenomenon, it forces us to investigate what is different about the molecules in each class that causes this behavior.

To answer this question, we might consider the molecular structure of molecules representative of each class. In doing so we would discover that the molecules belonging to the class containing water are *polar*. This simply means that the bonding of atoms together in these molecules results in an uneven distribution of electrical charge. One part of the molecule is slightly positively charged and another part is slightly negatively charged. On the other hand, the atoms of the molecules belonging to the class containing oil are bound together in a way that distributes the electrical charge fairly uniformly. These molecules are *nonpolar*. Figure 8.1 shows the molecular structure of water (with the charged parts indicated) and decane (a typical oil).

Interesting phenomena occur when a molecule is composed of two parts, one that by itself would mix with water and another that by itself would not. The part that would by itself be soluble in water is called the *hydrophilic* (water-loving) group and the part that would by itself not be soluble in water is called the *hydrophobic* (water-fearing) group. These molecules can display characteristics of both classes, so we call them *amphiphilic* molecules (loving both kinds). The two most important types of amphiphilic molecules are soaps and phospholipids, both of which will be discussed in detail later. Amphiphilic compounds are also known as *surfactants*. As we shall see, the reason for this is that amphiphilic molecules tend to migrate to the surface of a liquid. This fact has important scientific and technological implications.

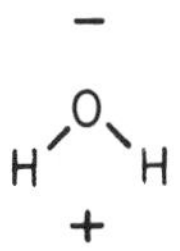

−

H_2O

+

(a)

CH_3 CH_2 CH_2 CH_2 CH_2 CH_2 CH_2 CH_2 CH_2 CH_3

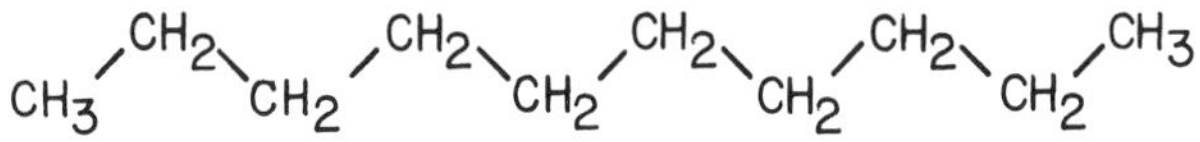

(b)

Fig. 8.1 A polar molecule, water (a), and a nonpolar molecule, decane (b).

A typical *soap* molecule is shown in figure 8.2(a). It consists of a polar head group formed by the bonding between a sodium atom and a carboxyl group, and a nonpolar group consisting of a hydrocarbon chain. A common symbol for this molecule is also shown. It consists of a small circle to represent the hydrophilic head group and a zig-zag line to represent the hydrophobic end chain. Soaps can be made using many other polar head groups together with hydrocarbon chains of various lengths. Some soaps have two hydrocarbon chains connected to the polar head group.

A typical *phospholipid* molecule is shown in figure 8.2(b). It consists of a larger polar head group containing a phosphorus atom and two nonpolar hydrocarbon end chains. Its simplified symbol is a small circle with two zig-zag lines attached. Again, many different phospholipids exist due to the large number of possible head groups and hydrocarbon chains of different lengths.

Structures Formed by Amphiphilic Molecules

If a small amount of amphiphilic material is mixed with water, it is possible for the molecules to go into solution. As the concentration of amphiphilic material increases, however, one of two possible structures begins to form. If the amphiphilic molecules have a strong polar head group relative to the nonpolar part of the molecule, the

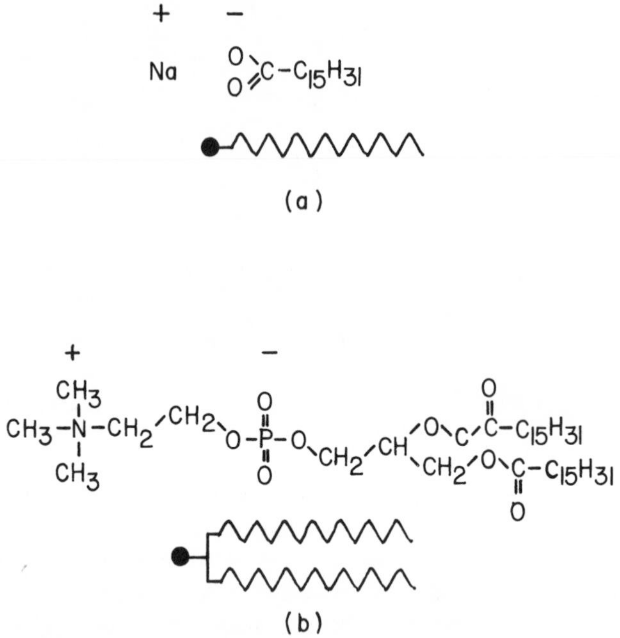

Fig. 8.2 A typical soap molecule, sodium laurate (a), and a typical phospholipid molecule, dipalmitoylphosphatidylcholine (b).

amphiphilic molecules begin to arrange themselves into spheres, with the polar head groups on the outside and the hydrocarbon end chains toward the center. This structure is called a *micelle* and is stable as long as the amount of amphiphilic material is above a certain concentration (called the *critical micelle concentration*). If the head group of the amphiphilic molecule is not strong relative to the hydrophobic part, the molecules begin to form spherical *vesicles*, in which double layers of amphiphilic molecules (called *bilayers*) form a shell with water on the inside and outside. Cross-sections of both of these structures are shown in figure 8.3.

It is not difficult to understand why both of these structures are stable. The hydrophilic head groups on the outside of the structures are in contact with the water molecules, while the hydrophobic end chains are in contact with each other and shielded from the water by the polar head groups. If the concentration of amphiphilic material is

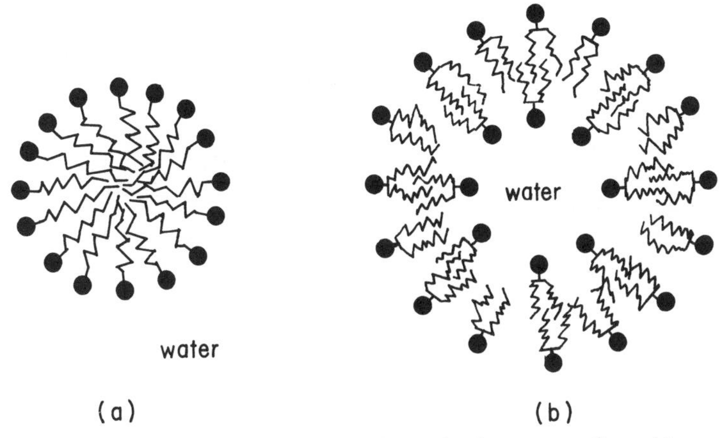

Fig. 8.3 Cross-sectional diagrams of two simple structures formed by amphiphilic molecules in water: (a) micelle and (b) vesicle.

increased further, more micelles or vesicles form. In some cases the size and shape of the micelles or vesicles remain fairly constant as their number increases. In other cases the shape of the micelles changes from spherical to cylindrical. In some cases vesicles with several bilayers form (each inside the next). The structure of these vesicles resembles an onion, but differs in that water occupies the region between each bilayer and the next.

It should be noted that similar structures begin to form if amphiphilic material is added to a nonpolar liquid such as oil. In this case the micelles or vesicles form with the polar head groups toward the inside and the nonpolar end chains toward the outside. Cross-sections of these so-called *inverted structures* are shown in figure 8.4. These structures can also change in shape or bilayer number as the amount of amphiphilic material is increased. Although I will concentrate my discussion on the structures formed in polar solvents such as water, keep in mind that corresponding inverted structures are also possible.

If the concentration of amphiphilic material is increased even more (usually in the vicinity of 50%), a point is reached where the micelles or vesicles combine to form larger structures. One such structure is called the *hexagonal phase* (sometimes called the *middle soap* phase), in which long cylindrical rods of amphiphilic molecules

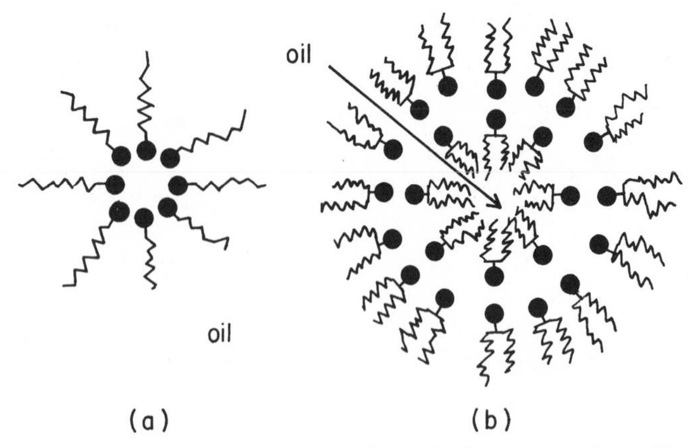

Fig. 8.4 Cross-sectional diagrams of two simple structures formed by amphiphilic molecules in oil: (a) micelle and (b) vesicle.

arrange the long axes of the rods in a hexagonal array. Another common structure that forms at even higher concentrations is the *lamellar phase* (sometimes called the *neat soap* phase), in which the amphiphilic molecules form flat bilayers separated from each other by water. One less common phase sometimes forms at concentrations between the hexagonal and lamellar phases. This *cubic phase* (sometimes called the *viscous isotropic* phase) is composed of spheres of amphiphilic molecules that arrange themselves in a cubic lattice. Whether the spheres are isolated from one another or connected in some way is at present an open question. Cross-sections of the hexagonal and lamellar phases are shown in figure 1.14. Undoubtedly, one of these structures is responsible for the "goo" in your soap dish.

These structures are liquid crystalline. The amphiphilic molecules diffuse throughout each of these structures, but in doing so orient themselves on average according to the structure. In some of these structures positional order is also present, so there are many similarities to smectic liquid crystals. Unlike smectic liquid crystals, however, the density of amphiphilic molecules varies drastically from point to point. The structures themselves are composed mostly of amphiphilic molecules while the water between the structures contains a relatively small number of single amphiphilic molecules. The amphiphilic molecules are free to diffuse throughout both the structures and the water. They do so maintaining both the order of the structures and the large concentration difference between the amphiphile-rich structures and the water-rich solvent.

There has been a great deal of research into the identification of which polar head groups cause the formation of these structures and which do not. From this work we have learned that the head group need not be ionic (formed by the ''donation'' of an electron by one atom to another) in order for these structures to form. Likewise, not all polar, nonionic head groups are effective in producing these structures. The formation of these liquid crystalline phases is obviously the result of a delicate balance between the hydrophilic and hydrophobic tendencies of the amphiphilic molecule, and thus may turn out to be a difficult process to fully understand.

You should not be surprised to find out that temperature has a significant effect on the stability of these phases. In fact, none of these phases can form if the temperature is not high enough. The reason for this is that all of these structures demand that the amphiphilic molecules be able to move relative to one another. If the temperature is too low, the molecules tend to form rigid crystalline structures. When mixed with water at this temperature, crystals of amphiphilic material simply exist in contact with the water. The temperature above which crystals do not form and these liquid crystalline structures do form is called the *Kraft temperature*. The Kraft temperature increases slightly as the concentration of amphiphilic molecules increases.

In order to summarize the behavior of these systems as both the concentration of amphiphilic material and the temperature is varied, we must resort to the same type of phase diagram used to describe how a mixture of two liquid crystals behaves. The concentration is plotted along the horizontal axis with increasing amphiphile concentration to the right. The temperature is plotted along the vertical axis. At any point on the diagram, one phase is stable and curves divide regions where different phases are stable. The phase diagram for a typical soap is shown in figure 8.5. Note the curves denoting the critical micelle concentration (the nearly vertical dashed line) and the Kraft temperature (the curve separating the crystal and water part from the rest of the diagram). The phase diagram for a typical phospholipid is shown in figure 8.6. The critical concentration for the

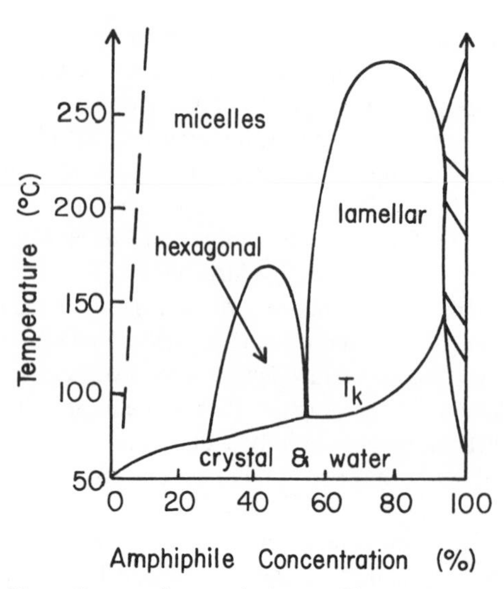

Fig. 8.5 Phase diagram for a typical soap. The nearly vertical dashed line shows the minimum concentration for micelle formation. The Kraft temperature T_K is the line separating the crystal/water part of the diagram from the rest of the diagram. Various liquid crystal phases occur in the region very close to the 100% concentration axis.

formation of vesicles is not shown on this diagram because it occurs at an extremely small concentration. Also, as explained in the next paragraph, the concept of the Kraft temperature is not nearly as useful for phospholipids. In both phase diagrams, additional liquid crystal phases (usually inverted phases) exist at very high concentrations and various crystalline phases form at low temperatures.

The *gel* phases of certain phospholipids are quite interesting. The amphiphilic molecules form bilayers much like the lamellar phase, but the hydrocarbon end chains are rigid, much as they are in the solid phase. This is unlike the lamellar phase, in which the hydrocarbon end chains are free to adopt changing configurations, much as they are in the liquid phase. The rigid end chains of the gel phase allow the phospholipid molecules to arrange their polar head groups in a closely-spaced hexagonal array. Many substances have more than one gel phase, with the orientational and positional order increasing

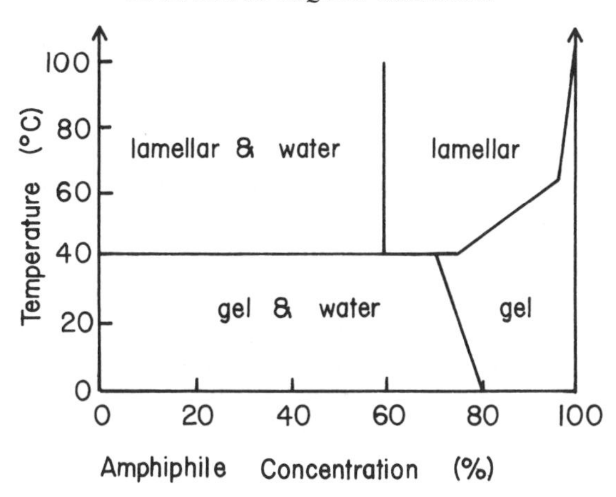

Fig. 8.6 Phase diagram for a typical phospholipid. Various liquid crystal and solid phases occur at high concentration and have been omitted from the figure.

with each phase change as the temperature is lowered. Since no drastic change between liquid crystalline micelles and solid crystals mixed with water takes place, defining one of these transition temperatures as the Kraft temperature is problematic. Finally, it should be pointed out that the bilayers are rippled in some gel phases. This is not true of the lamellar phase.

The Miscibility Gap

Nothing underscores the delicate balance between opposing tendencies that takes place in the formation of these structures like the behavior of certain nonionic amphiphiles. Just as their ionic counterparts, these amphiphilic compounds first form micelles and then the various liquid crystal phases as the concentration of amphiphilic material increases. However, if the temperature is high enough, the sequence is quite different as the concentration of amphiphile increases. At very low concentrations micelles form as usual, but at a certain concentration they break up and the system forms two separate phases, one rich in water and the other rich in amphiphile. If the concentration is further increased, the micelles again form and the two separate phases combine to form the usual homogeneous phase of water and micelles.

At these higher temperatures, therefore, a region exists where the two compounds cannot be mixed (that is, they are not *miscible*), so researchers refer to this as the *miscibility gap*. This gap can be seen in the phase diagram of the nonionic amphiphile/water system shown in figure 8.7. The miscibility gap is the region labeled "two phases" in the figure. Note that the miscibility gap widens as the temperature increases. In some cases even this feature reverses itself, with the miscibility gap becoming narrower and even disappearing if the temperature is raised sufficiently high. This bizarre behavior is direct testimony that the formation of micelles in amphiphilic/water systems is the result of a delicate balance. Changes in temperature and concentration are capable of drastically changing the balance condition.

Amphiphilic Compounds in Water/Oil Mixtures

The situation is even more interesting if oil is added to the amphiphile/water mixture. Although oil normally is not miscible with wa-

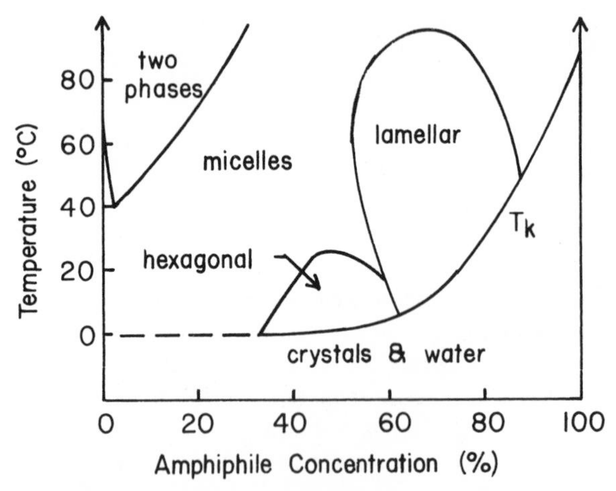

Fig. 8.7 Phase diagram for dodecyldimethylphosphide oxide in water. The region labeled two phases shows the miscibility gap of this system.

ter, the amphiphile/water mixture has little difficulty in incorporating it into its structure. Let us concentrate on the micelle phase of the amphiphile/water mixture. As oil is added, it accumulates inside the micelles in contact with the *oliophilic* (oil-loving) hydrocarbon chains of the amphiphilic molecules. Because no separate oil-rich phase forms, the amphiphile/water mixture effectively "dissolves" the oil. This process is why soaps and detergents are effective as cleansers. We can describe this phenomenon in a slightly different way. Imagine placing water and oil together in a glass. Shake or stir as you might, the two do not mix and eventually separate into two phases. Now imagine adding some amphiphilic material to the water and oil. After a good shake the liquid in the glass appears slightly cloudy but homogeneous, and it remains that way. The amphiphilic molecules have migrated to all the boundaries between the two liquids (it is a surfactant) and have effectively "shielded" one from the other. So water and oil can in fact mix if some amphiphile is added.

Let us consider this process in more detail. If we start with a mixture of amphiphile and water in the micelle phase, the addition of a small amount of oil causes the micelles to swell slightly as oil is incorporated into their centers. These swollen micelles are completely stable. As more oil is added, the micelles continue to swell and eventually reach the point where the amphiphilic molecules actually are the barrier between one phase (oil) and another phase (water). At this point the system is an *emulsion*, in that it must be described as a combination of two phases. Amphiphilic compounds are therefore *emulsifiers*. For oil concentrations less than that necessary to form an emulsion, we call the single phase of slightly swollen micelles a *microemulsion*.

To partially describe what phases result from the mixture of three components, we can use the same type of phase diagram as before. The only problem is that we must select the ratio of oil to water and keep it fixed. We can then add amphiphilic material to the selected oil/water mixture and determine what phases are present at various temperatures and amphiphile concentrations. A representative diagram is shown in figure 8.8. Notice that in addition to lamellar and micelle phases, certain concentrations cause the oil and water to separate into either two phases (oil and water) or three phases (oil, water, and amphiphile).

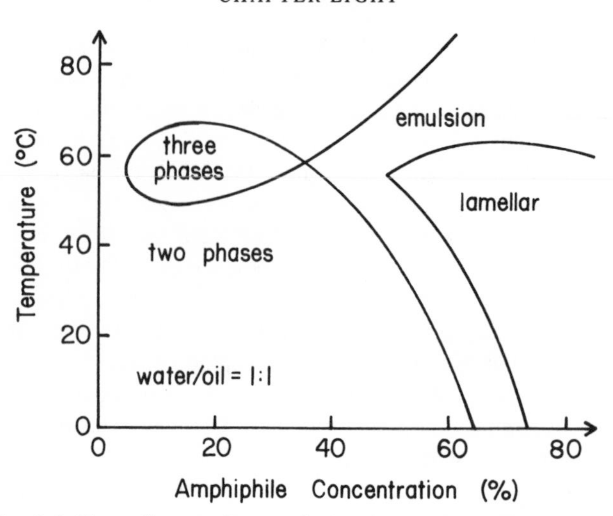

Fig. 8.8 Phase diagram for a typical tertiary system. The amount of water and oil is always equal, but the amount of amphiphile varies.

Another way to partially describe a three-component mixture is to keep the temperature constant and vary the relative amounts of all three components. All possible ratios of the three components can be represented as points inside an equilateral triangle, where each vertex of the triangle represents 100% of one of the components. To determine the amount of component represented by a point inside the triangle, simply drop a line perpendicular to the side opposite from the vertex representing 100% of that compound. The length of this line is a measure of the concentration of that compound on a scale where the height of the triangle is equal to 100%. Figure 8.9 shows a representative phase diagram. Note the liquid crystalline and micelle phases, as well as the emulsion and two-phase regions.

Applications

Our discussion of amphiphile/water/oil mixtures leads to the oldest and most important application of lyotropic liquid crystals—their use as detergents. Amphiphilic compounds have been used as soaps for over three thousand years, yet our knowledge of how they work is very recent. Many questions remain unanswered. For example, the

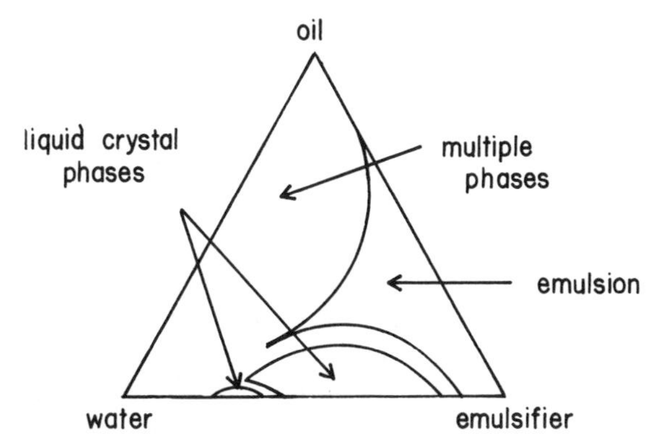

Fig. 8.9 Phase diagram for a typical water-oil-emulsifier system. See text for an explanation of how to read a tertiary phase diagram.

phase diagrams such as the one in figure 8.8 show that certain amphiphiles are most effective in making oil soluble in water at a particular temperature. We do not completely understand why this is so. There can be no doubt, however, that this question touches an important aspect of technology. Just think of how useful it would be to develop detergents that work best at the temperature they encounter during a specific application. Perhaps a laundry detergent could be developed that really does wash in cold water as effectively as our modern detergents do at hot water temperatures. Needless to say, the detergent industry is huge, producing soaps, powders, creams, and foaming agents of all kinds.

A good deal of the research into amphiphile/water/oil systems is being done for the crude oil industry. A large portion of the oil trapped in porous rocks (up to 50%) remains there after an oil well runs out. If these oil fields could be flooded with a cheap amphiphile/water mixture that effectively incorporated the oil into its micelles, then the oil could be recovered by pumping out the amphiphile/water/oil mixture from the well. The amount of oil potentially available through such a process is huge.

The food industry is an active user of lyotropic liquid crystals as food emulsifiers. The addition of these compounds serves to maintain texture, color, flavor, or viscosity. Emulsifiers are used in mayonnaise, salad dressings, marshmallow, whipped cream, beer, cheese, ice cream, and jelly. One interesting example concerns the production of bread. Good mixing of the ingredients in bread is necessary in order for the bread to rise and bake properly. Some grains contain these liquid crystalline compounds naturally, so their action as emulsifiers guarantees a good mixture of the ingredients. Grains that lack these natural emulsifiers and normally produce inferior bread can be made to produce good bread by the addition of synthetic liquid crystals.

There may be some significant medical applications for lyotropic liquid crystal systems. For instance, some drugs are not very soluble in blood and consequently do not get carried by the bloodstream to all the important active sites of the body. Addition of a small amount of amphiphilic material to the blood could make the drug more "soluble" and therefore more efficiently transported throughout the body. The amphiphile would have to be nontoxic of course, but if it could be designed to work best at body temperature, only a small amount need be added to the bloodstream.

Another possible medical use is only slightly different. Drugs that cannot now be taken orally because the enzymes of the digestive tract break them down might be encapsulated by an amphiphilic compound that would protect the drug while in the digestive tract. The amphiphilic compound could be designed to dissolve in the bloodstream, thus releasing the drug.

Thermotropic-like Liquid Crystals Formed by Micelles

Certain tertiary mixtures of a highly polar liquid, a slightly polar liquid, and an amphiphile form micelles or vesicles that are not spherical in shape. Usually this happens only if the concentration of amphiphilic compound is within a very narrow range. Within this range, rod-shaped micelles, disk-shaped micelles, or micelles with an elliptical cross-sectional area form. These micelles behave much like typical liquid crystal molecules, spontaneously orienting themselves with their axes along a preferred direction. Studies have uncovered examples where the micelles orient to form just about all of the basic liquid crystal phases. The narrowness of this region indicates that this behavior is the result of a very delicate balance that sometimes causes nonspherical micelles to form. The phase diagram for one such mixture is shown in figure 8.10. Notice how small the range of concentration in the figure is.

One Final Observation

The amount of research performed on lyotropic liquid crystals over the last twenty-five years is small when compared to the amount done on thermotropic liquid crystals. However, there are indications that this is beginning to change. As we have seen, the potential applications of lyotropic liquid crystals lie in four important areas: detergents, food emulsifiers, oil recovery, and medical technology. The amount of money at stake for companies or individuals who successfully harness these applications is staggering. In the past, such a situation in other areas of science has always resulted in a dramatic increase in research activity. There is every reason to suspect that the same will be true for lyotropic liquid crystals.

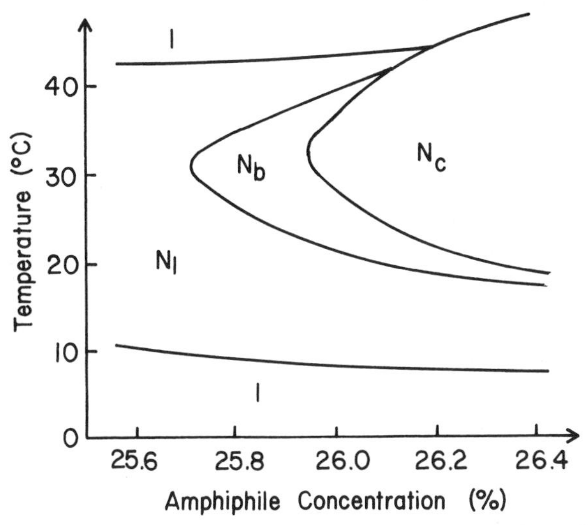

Fig. 8.10 Part of the phase diagram for the heavy water-decanol-cesium laurate system. The micelles are either rodlike (N_c), disklike (N_l), or more complicated in shape (N_b).

CHAPTER 9

Polymer Liquid Crystals

The chemistry and physics of polymers is a new and exciting area of science, which has already made a significant impact on our way of life. The future is even brighter; polymer science is an established and rapidly growing field of both basic science and engineering. As we shall see in this chapter, polymer liquid crystals continue to play an important role in these developments. Although it is beyond the scope of this book to include a full discussion of the entire field of polymers, we must begin with a general introduction to polymers, emphasizing those aspects important for understanding polymer liquid crystals. The different types of polymer liquid crystals and their properties are then discussed. The chapter ends with a description of a few areas where polymer liquid crystals have important applications.

In some sense, a more appropriate title for this chapter is "macromolecular liquid crystals." A *macromolecule* is simply a very large molecule, and as we shall see, polymers are a specific type of macromolecule. Since some of the discussion toward the end of the chapter concerns very large molecules that are not polymers in the traditional sense, the more general term (macromolecular liquid crystals) might seem to be a better choice. However, the advances in knowledge and the development of new technology in polymer liquid crystals over the last twenty years have been so important that it is indeed appropriate that they receive top billing.

POLYMERS

Any large molecule composed of a long sequence of repeating units is called a *polymer*. The unit that is the basic building block and that is repeated in the molecule is called the *monomer*. The monomer itself consists of atoms with chemical bonds between them; likewise, chemical bonds connect the monomers together in the polymer. The chemical reaction in which these bonds between monomers form is called the *polymerization* process. If we represent the monomer with the letter *M*, then we can represent the polymer by a series of such letters, *MMMMMMMMMM*. Figure 9.1 contains a typical example. The basic unit comes from the ethylene molecule, which is shown on the left. Part of the polymer (called polyethylene) is shown on the right, along with the symbol that we use to denote the polymer. The small letter *n* signifies that the unit between the brackets is repeated many times in the polymer.

The monomer can consist of just a few atoms, as in the case of polyethylene, or its structure can be quite complex. The properties of the monomer ultimately determine the properties of the polymer, and we shall see that a huge range of physical properties is possible. Figure 9.2 contains some examples of polymers with which you are familiar. As you look at some of these examples, consider how different the properties of these materials are. The number of possible monomers is extremely large, which means that many different polymers can be produced.

The polymerization process rarely creates polymer molecules that have the same number of monomers. Therefore, any sample of the polymer material contains polymer molecules made from different numbers of monomers. To describe a polymer sample, we must state the average number of monomers in a polymer molecule (called the *degree of polymerization*), and by how much the majority of the polymer molecules differ from this average number. For example, a sam-

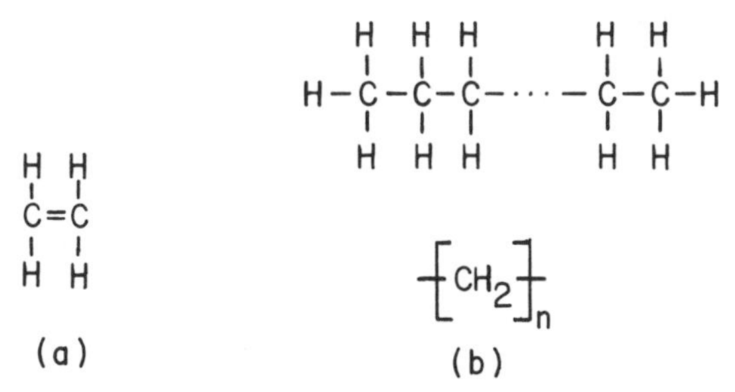

Fig. 9.1 Representation of a typical polymer. The ethylene monomer is shown in (a) and two representations (the full polymer and the repeating entity) of polyethylene are shown in (b).

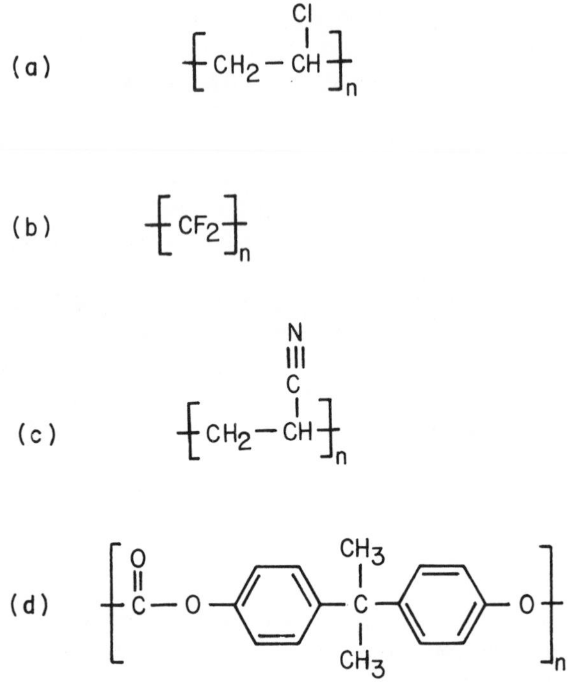

(a) $\left[CH_2 - CH(Cl) \right]_n$

(b) $\left[CF_2 \right]_n$

(c) $\left[CH_2 - CH(C \equiv N) \right]_n$

(d) $\left[C(=O) - O - C_6H_4 - C(CH_3)_2 - C_6H_4 - O \right]_n$

Fig. 9.2 Four well-known polymers: (a) polyvinylchloride (PVC); (b) Teflon; (c) Acrylan; and (d) Lexan.

ple of polymer might have a degree of polymerization equal to 100, with roughly two-thirds of the polymer molecules having a number of monomers differing from this average by less than 15 (that is, between 85 and 115). In general, it is quite difficult to force a polymerization reaction to produce polymers all with the same number of monomers, because the linking of monomers occurs quite randomly during the reaction. Thus there is little control over the length of each polymer molecule. The result is usually a statistical distribution in the number of monomers per polymer.

Polymers can also be made from a chemical reaction in a mixture of two types of monomers. The result of this process is called a *copolymer*. If the two types of monomers (*M* and *m*) combine at random to form the polymer, a *random copolymer* results (*MmMMm-*

MmmmMmMM). If the two monomers form short sequences of one type first (*MMMM* or *mmmmm*), which then combine to form the final polymer (*MMMMmmmmMMMMMmmmm*), a *block copolymer* results. Finally, if short sequences of one monomer (*mmmmm*) are attached as side chains to a very long sequence of the other monomer (*MMMMMMMMM*), a *graft copolymer* is formed. The properties of copolymers depend on many factors. These include the type of copolymer formed, the characteristics of the two monomers, the amount of one type of monomer versus the other, and the size of the short sequences in block and graft copolymers. Again, a huge number of possible monomers exist, so the combinations that can be chosen are even greater. In addition, the different ways the two monomers can be combined into a copolymer make the number of possible copolymers virtually limitless. The range of physical and chemical properties of all these copolymers is almost unimaginable.

As is true of all substances, polymers can exist in more than one phase. The solid phase of polymers can be either crystalline or amorphous. In the crystal phase, the polymer molecules occupy positions in a specific arrangement. In the amorphous phase, the polymers are arranged quite randomly but are not free to move to another position. The amorphous solid phase of polymers is also called the *glassy* phase, since glass possesses an amorphous structure. Often these two phases can coexist over a significant temperature range. A solid polymer, whether crystalline or glassy, will melt when heated. If it melts to a liquid crystal phase of some sort, it is called a polymer liquid crystal. In many cases the polymer melts to form a liquid, without forming a liquid crystal phase. At high enough temperatures, a gas phase might be possible, but the polymer usually decomposes before this temperature is reached.

One additional comment should be made about these phase transitions in polymers. Because a typical polymer sample contains polymer molecules with varying numbers of monomers, the temperature at which a phase transition takes place is usually not as precisely defined as in materials consisting of identical molecules. The reason for this is not difficult to understand. The exact temperature at which a polymer changes phase depends on the number of monomers in it. If the sample contains molecules with different numbers of monomers, all molecules do not behave identically when the temperature is changed. The result is a phase transition that is spread over a range of temperatures, as opposed to a pure material where the phase transition takes place at one temperature. This phenomenon is quite general; mixtures of thermotropic liquid crystals show the same effect. How wide the temperature interval of the phase transition is depends on many factors, including the components of the mixture and the relative amounts of each component. The most important factor in the case of polymers is the degree to which the number of monomers in the polymer molecules varies within the sample.

Thermotropic Polymer Liquid Crystals

If polymers are formed from certain types of monomers, the polymer possesses liquid crystal phases in addition to the solid and liquid phases. Not surprisingly, these monomers closely resemble molecules that by themselves form liquid crystals, so they fall into two general classes. The first class consists of monomers that are fairly rigid, anisotropic, and highly polarizable. Just as in thermotropic liquid crystals, these monomers are either rod-shaped or disk-shaped. The second class of monomers causing liquid crystallinity in the polymer are amphiphilic monomers. Very little is known about the behavior of polymers formed from amphiphilic monomers, mainly because little work has been done in this area. Therefore, this chapter deals almost exclusively with polymer liquid crystals formed from rigid, anisotropic, nonamphiphilic monomers.

The monomers can be attached together to form a polymer in two different ways. If the monomers form a long single chain by attaching to one another, a *main chain polymer* results. If the monomers end up as branches extending away from the polymer chain, then a *side chain polymer* forms. In either case, the attachment to the polymer chain can be along either a short or long axis of the monomer. Diagrams of these four types of polymers are shown in figure 9.3, using both rods and disks to represent the monomers. How chemists get monomers to attach themselves in these various ways is an important question, which will be addressed when we discuss the synthesis of polymer liquid crystals.

The use of rigid and polarizable monomers sometimes makes things quite difficult for the polymer scientist. In some cases using

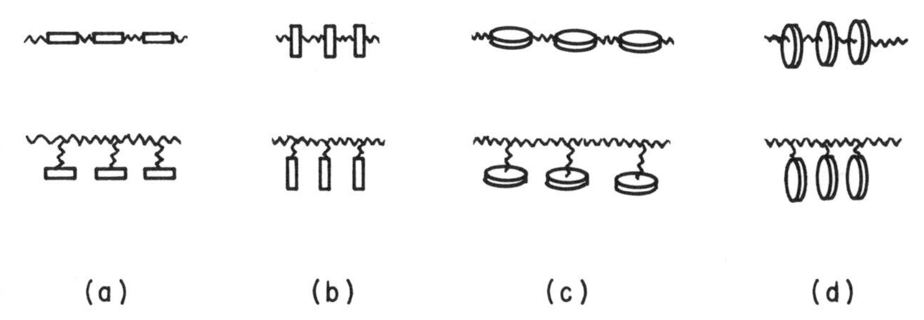

Fig. 9.3 Types of polymers that can be formed with anisotropic units (shown as rods or disks). The top row are all main chain polymers and the bottom row are all side chain polymers.

such monomers produces polymers that melt at high temperatures, making the study of its liquid crystalline or isotropic phases quite difficult. In addition, polymers made from such monomers are sometimes fairly insoluble in the normal solvents. Again, this makes working with these polymers more difficult. In spite of these hardships, scientists have created many useful polymer liquid crystals and have uncovered general rules that help them create materials with the properties they desire.

For example, a great deal of experimentation has shown that useful polymer liquid crystals result when the monomer consists of a rigid section away from the point of attachment and a more flexible section near the point of attachment. The polymers shown in figure 9.3 are drawn using such monomers. The rod or disk represents the rigid part of the monomer and the zig-zag line represents the more flexible part. The flexible part near the point of attachment is called the *spacer*. Numerous studies have been carried out to investigate what effect the length of the spacer has in a polymer when the rigid, anisotropic section remains the same. Likewise, many other studies have kept the length of the spacer constant and have used various rigid groups to make different polymers. All of this work has uncovered hundreds of polymer liquid crystals with nematic, chiral nematic, and smectic phases.

Whether main chain or side chain polymers, it is the rigid, anisotropic section of the monomers that display orientational order in the liquid crystal phases. In the nematic phase, these rigid sections tend to point along a preferred direction just as in thermotropic liquid crystals. There is no positional order and the other parts of the polymer display no orientational or positional order. Figure 1.13 shows the type of order present in nematic polymer liquid crystals for both main chain and side chain polymers. A rod has been used to represent the rigid, anisotropic part of the monomers. The same order parameter S, used to describe the amount of orientational order in liquid crystals, is also used for polymer liquid crystals. Its value describes how well the rigid, anisotropic parts of the polymer are ordered. Determinations of S for polymers reveal that the amount of order in polymer liquid crystals is close to that in liquid crystals. As can be seen by comparing figures 9.4 and 1.5, in some cases even the temperature dependence is very similar to the behavior in thermotropic liquid crystals.

In some cases, the preferred direction in a polymer liquid crystal is not constant but rotates in helical fashion just as it does in a chiral nematic liquid crystal. As you might expect, monomers that resemble nematic liquid crystal molecules produce nematic phases in the polymer, while monomers that are similar to chiral nematic liquid crystal molecules form chiral nematic phases in the polymer. Finally, it is possible for the rigid, anisotropic sections of the monomer to tend to

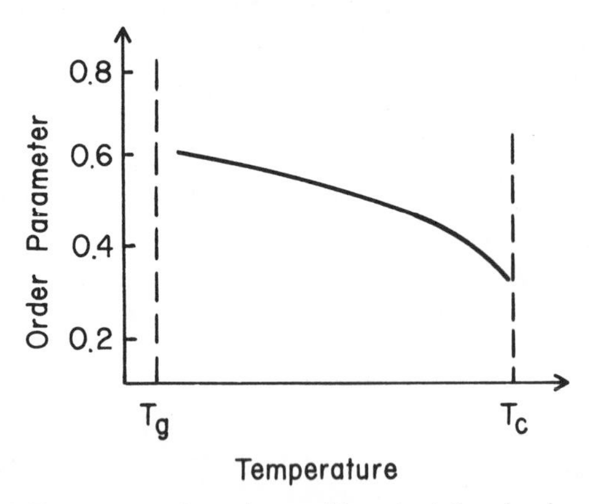

Fig. 9.4 Temperature dependence of the orientational order parameter in a typical polymer liquid crystal. T_c is the transition temperature to the isotropic phase and T_g is the transition temperature to the glassy phase.

position themselves in layers as they orient in the liquid crystal phase. In this case a smectic liquid crystal phase is present in the polymer. Keep in mind in all of these cases, that it is only the rigid, anisotropic part of the polymer that possesses order of any kind. The more flexible parts of the polymer, whether main chain or side chain, wander throughout the phase displaying little or no order of any kind. This fact is also evident in figure 1.5.

A polymer liquid crystal may possess one or more liquid crystal phases. Figure 9.5 contains an example of a polymer liquid crystal and the phases it forms. As is true for liquid crystals, many sequences of phases are possible. Because pure samples of these polymers form a liquid crystal phase, they are called *thermotropic* polymer liquid crystals.

There is one other example of a thermotropic polymer liquid crystal that is quite different from the ones I have been discussing. Certain petroleum pitches and coal tars contain hydrocarbon molecules consisting of fused aromatic rings. When these materials are heated, the molecules polymerize into even larger molecules consisting of fused aromatic rings. Examples of both a smaller "monomer" and a

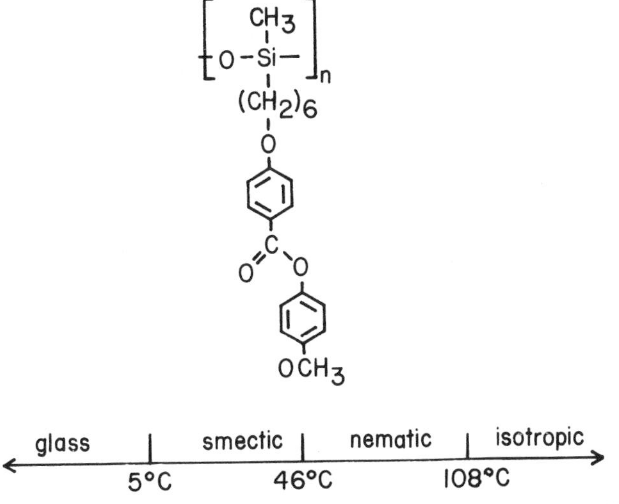

Fig. 9.5 Structure and phase diagram for a typical polysiloxane side chain polymer.

larger "polymer" are shown in figure 9.6. Whereas we sometimes refer to the chainlike polymers as *linear polymers*, these large disk-like molecules are called *network polymers*. The large disklike networks tend to orient with their planes parallel to one another, just as is the case for the molecules of a discotic liquid crystal.

Thermotropic polymer liquid crystals form when the polymeric material is heated to a point where the solid phase melts. For this reason, these liquid crystal phases are sometimes referred to as *polymer melts*. There is one other possibility. Some very large macromolecules (often polymers), form phases with orientational order when dissolved in ordinary solvents. In these cases, concentration is more important than temperature, so these phases are called either *lyotropic polymer liquid crystals* or *polymer solutions*.

Lyotropic Polymer Liquid Crystals

The first requirement for a macromolecule to form a liquid crystal phase in solution is that the molecule be fairly rigid. Rigid monomers that form strong bonds with each other represent one possibility. Macromolecules that adopt a single- or double-stranded helical structure with strong forces holding the structure in place also act as very rigid molecules and form liquid crystal phases. The second requirement is that the macromolecules must dissolve in the solvent. Liquid crystalline phases only form when the concentration is high enough so that these macromolecules constantly interact with one another. This cannot be achieved if the macromolecules do not dissolve in the

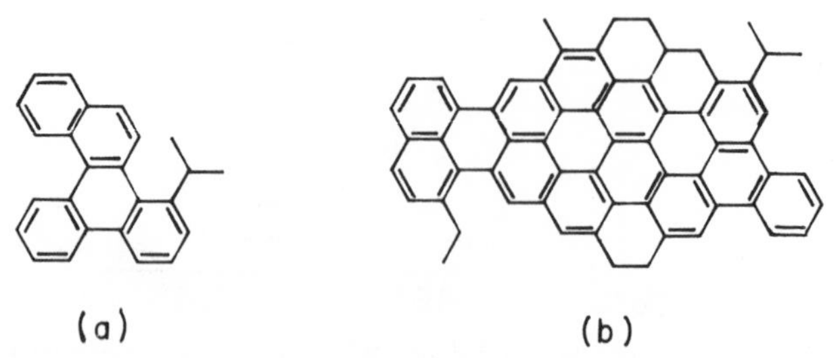

Fig. 9.6 Example of a network "monomer" (a) and "polymer" (b).

solvent to high enough concentrations. As previously mentioned, these two requirements are often mutually exclusive. The more rigid a molecule is, the less soluble it is. Scientists sometimes must use very strong solvents (like sulfuric acid) to dissolve enough material. Still, many synthetic and natural large molecules form liquid crystal phases in many solvents.

A good example of a class of synthetic polymer liquid crystals is known as the *polyamides*. Three typical monomers from this class are shown in figure 9.7. Notice that the monomer and the bonds between monomers are similar to what one finds in the central, rigid part of some liquid crystal molecules. In both cases the resulting molecule or polymer is fairly rigid.

Just about all of the large molecules that form liquid crystal phases due to a helical structure are biologically important molecules. For example, many derivatives of *cellulose* are polymer liquid crystals.

(a) (b) (c)

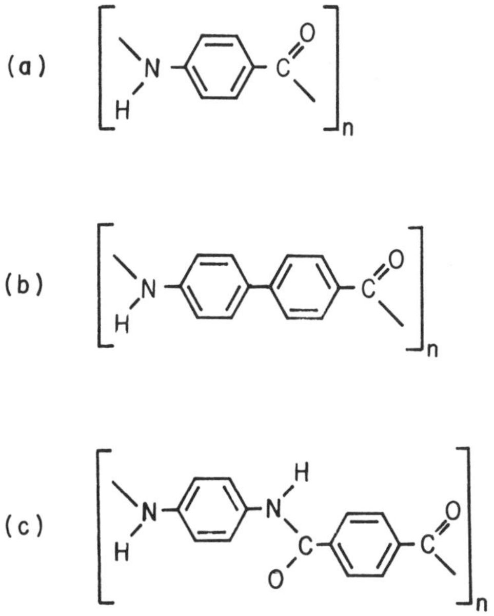

Fig. 9.7 Three examples of aromatic polyamides with liquid crystal phases. The polymer shown in (c) is known as Kevlar and is used to produce high-strength fibers.

Cellulose itself is a polymer consisting of a linear chain of repeating units. The unit is derived from *glucose*, a sugar, and is shown in figure 9.8. Notice that there are chemical entities (−H, −OH, and $-CH_2OH$) attached to the central part of the monomer. If some of these are replaced with larger chemical entities, the linear chain adopts a helical structure, gaining rigidity and producing liquid crystal phases.

Amino acids form the basic building block for the next type of polymer liquid crystal. As shown in figure 9.9, a typical amino acid possesses two quite different reactive groups. The carboxyl group

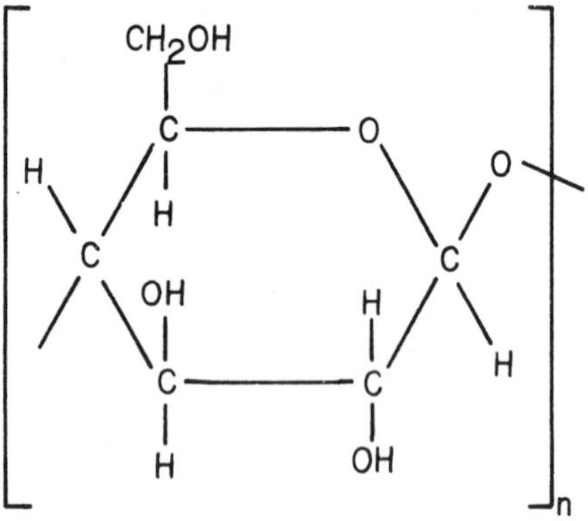

Fig. 9.8 The basic unit of cellulose, which is derived from the sugar glucose.

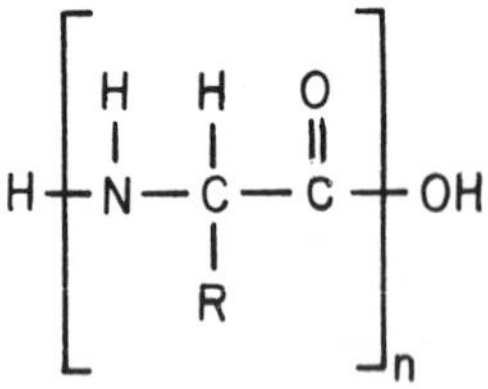

R = hydrogen or organic group

Fig. 9.9 The basic unit of a polypeptide. Each unit is derived from an amino acid.

(−COOH) on the end of one amino acid can react with the amino group (−NH_2) on the end of another amino acid to form a chain of amino acids. The resulting macromolecules are called *polypeptides*. Their structure is basically a long helical chain, which causes the macromolecule to be fairly rigid. When dissolved in water at high enough concentrations, liquid crystal phases occur.

Some polynucleic acids are also polymer liquid crystals. In these macromolecules, two chains form a double helix. The chains are composed of alternating sugar (like cellulose) and phosphate (like phospholipid) residues. The two helical chains are attached together by nitrogenous base pairs bound to the sugar residues in the chains. Deoxyribonucleic acid (DNA) is the most important example. The double helical structure is very rigid and these polynucleic acids form liquid crystal phases at high concentrations in water.

The last example of a macromolecule that forms a liquid crystal phase is a linear virus such as tobacco mosaic virus. Linear viruses consist of proteins (chains of linked amino acids) around a small amount of RNA (ribonucleic acid) or DNA. The structure is again a helix and the virus is rigid enough to form liquid crystal phases.

Because all of these macromolecules have helical structures, they interact in a fashion which prefers that the molecules not be exactly parallel to one another. As a result, some of these molecules form chiral nematic phases instead of nematic phases. Some even produce vivid colors due to selective reflection. One interesting feature, however, is that the pitch of these chiral nematic phases depends not only on the polymer but also on the solvent. In some solvents a certain polymer will form a right-handed chiral nematic phase and in another solvent it will form a left-handed phase. In some cases, the proper concentration in the proper solvent will produce a nematic phase. Obviously, the forces between these large molecules are due to the helical molecules themselves and to the interactions between these large molecules and the solvent molecules.

Synthesis of Polymer Liquid Crystals

If two groups on the monomer can react to form chemical bonds between the two different groups, then a method called *condensation* can be used to synthesize polymers. For example, consider a molecule with a carboxyl group (−COOH) on both ends, and another molecule with a hydroxyl group (−OH) on both ends. These two molecules can bind together to form another molecule, releasing water during the reaction. This new molecule, with a carboxyl group at one end and a hydroxyl group at the other end, can react with the other similar molecules to form a polyester polymer. In a slightly different example, one molecule with carboxyl groups at both ends reacts with another molecule with amine end groups ($-NH_2$), giving off water. The resulting molecule has a carboxyl group at one end and an amine group at the other end. These can polymerize with other similar molecules to form polyamides. Nylon can be synthesized in this way and, as previously mentioned, amino acids can form polymers in a similar fashion. Both of these condensation examples are shown in figure 9.10.

It is possible to synthesize polymers without releasing a product

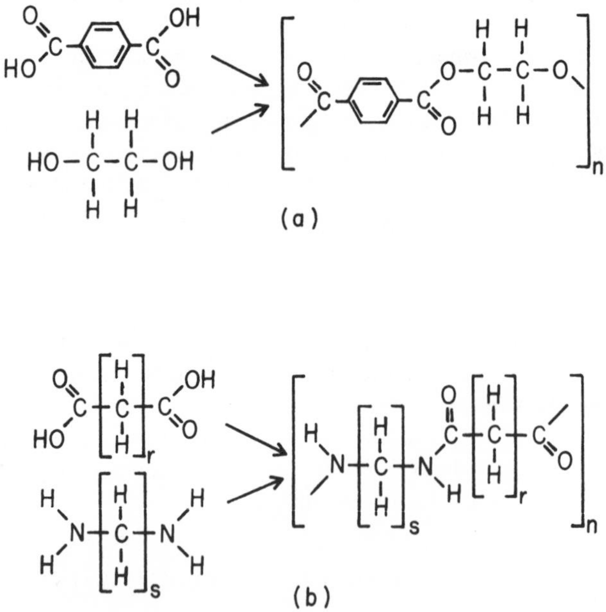

Fig. 9.10 Two examples of condensation reactions: (a) terephthalic acid and ethylene glycol form the polyester Dacron; (b) two monomers (*r* and *s* give the number of carbon atoms) form a polymer.

such as water. This works effectively, for example, if the monomer originally contains an unreactive electron pair that can be made available for bonding. Consider the propylene monomer shown in figure 9.11(a). Reaction with a molecule with one unpaired electron (shown as *R* in the figure) can cause one of the electrons in the carbon-carbon double bond to pair with the unpaired electron of the molecule. As evident from figure 9.11(b), the new molecule formed in this reaction still has one unpaired electron. This new molecule can react with other monomers with a double bond, just as the first molecule with one unpaired electron did. This is shown in figure 9.11(c). In this way, the polymer grows longer as more monomers are added on. In the case of propylene monomers, the result is the polymer polypropylene. This type of polymerization reaction is called *addition*, and is very important in the production of well-known polymers such as polyvinylchloride (PVC) and tetrafluoroethylene (Teflon).

One final example of a technique for synthesizing polymer liquid crystals is quite different. Instead of simply starting with monomers, one reacts monomers with a polymer. This method works well for the synthesis of side chain polymers, where the starting materials are a long flexible polymer and a short rigid monomer. Quite simply, rigid monomers bind to the long flexible polymer to form a side chain polymer. A good example of such a *modification* reaction is the syn-

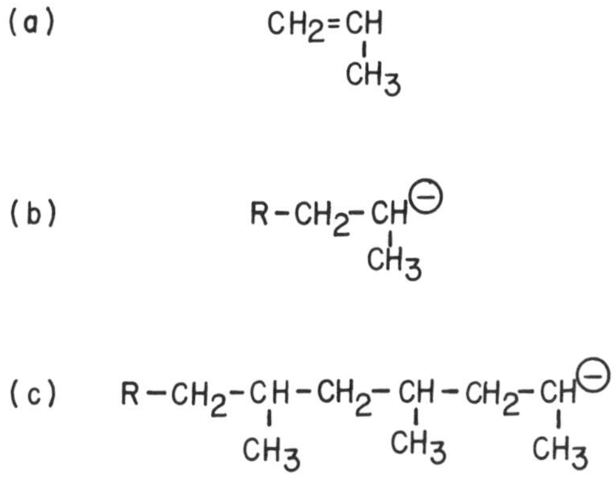

(a) $CH_2{=}CH(CH_3)$

(b) $R{-}CH_2{-}CH^{\ominus}(CH_3)$

(c) $R{-}CH_2{-}CH(CH_3){-}CH_2{-}CH(CH_3){-}CH_2{-}CH^{\ominus}(CH_3)$

Fig. 9.11 An example of an addition reaction: (a) propylene monomer; (b) result of initial reaction with a free radical *R*; (c) result of two additional reactions in the formation of polypropylene.

thesis of the polysiloxanes. Figure 9.12(a) shows the repeating unit of a long flexible polymer and figure 9.12(b) shows a rigid monomer with a reactive end group. The modified side chain polymer is shown in figure 9.12(c).

It is important to keep in mind that these polymerization reactions can proceed in a variety of ways. In some cases, the number of monomers in a typical polymer molecule increases as the reaction takes place, but in such a way that most of the polymer molecules have nearly the same degree of polymerization. The distribution of the number of monomers in a typical polymer molecule therefore stays fairly sharp during the reaction, but this distribution constantly shifts to higher numbers of monomers per polymer molecule. In other cases, the polymer molecules with the highest degree of polymerization react with each other rather than monomer molecules, quickly producing a small number of very large polymer molecules. The dis-

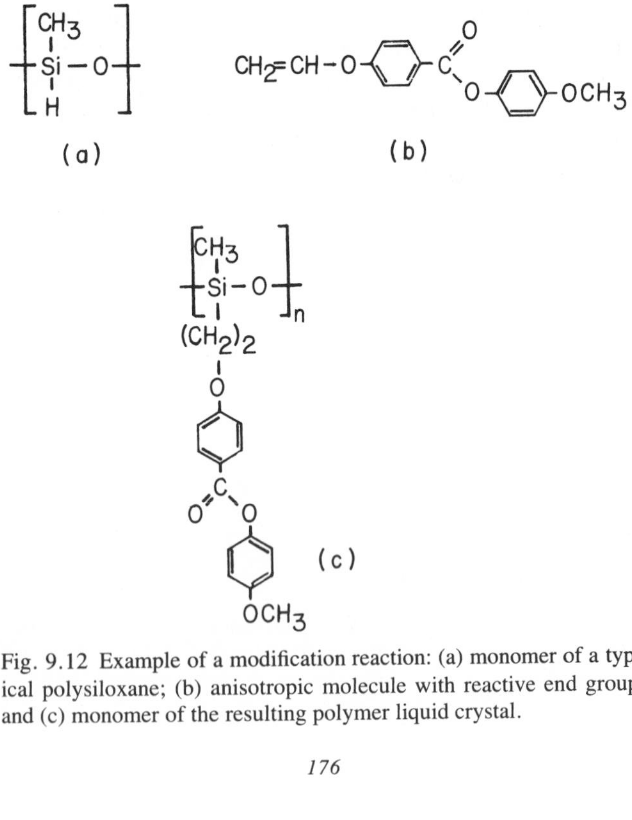

Fig. 9.12 Example of a modification reaction: (a) monomer of a typical polysiloxane; (b) anisotropic molecule with reactive end group; and (c) monomer of the resulting polymer liquid crystal.

tribution of the number of monomers per polymer molecule is quite different in this case. As soon as the reaction begins, there are some polymer molecules with a large number of monomers while the rest of the polymer molecules have just a few monomers. As the reaction proceeds, the number of very large polymer molecules increases while the number of very small polymer molecules decreases. There is no gradual shift of the distribution from lower to higher numbers of monomers, simply a gradual transfer from a majority of small polymer molecules to a majority of large polymer molecules.

How these polymerization reactions cease is also important. When the number of reactive end groups decreases to the point where there is little chance for further reactions between them, the polymerization process stops. Hopefully, this occurs because most of the reactive end groups have combined to form large polymer molecules. There are other possibilities, however. A reactive end group can bind to an impurity or to an unreacted starting group. Alternatively, a reactive end group can bind to a normally unreactive group (usually through the transfer of an electron). The point is that there are some reactions that continue the production of longer polymers and other reactions that terminate the production of such polymers. As the process proceeds, both types of reactions take place, until the number of reactive end groups is small. The result of a polymerization process therefore depends on the relative speeds of the two types of reactions during the process. With this in mind, it's easy to understand why these polymerization processes normally result in polymers of very different chain lengths. Extreme care is required to synthesize polymers containing roughly the same number of monomer units.

Applications of Polymer Liquid Crystals

Before we can appreciate one of the most important uses of polymer liquid crystals, we must first recognize why polymers in general are such useful materials. If a polymer in the liquid state is allowed to cool into the crystalline or glassy state, it possesses properties similar to the solid phases of other large organic molecules. Generally, these properties are not all that useful for applications. However, if the molten polymer solidifies while it is flowing or when it has just stopped flowing, the long and normally tangled polymer molecules are stretched out and oriented in the solid phase. The properties of this material are very different and extremely useful for many varied applications. In other words, the usefulness of polymers depends on our ability to orient the long polymer chains. Since the polymer chains in the liquid crystal phases of certain materials are already ordered, these materials can be solidified with even greater ordering of the polymer chains. This fact leads to one of the most important applications of polymer liquid crystals: ultra-high-strength fibers.

All processes that utilize flow to orient polymer molecules involve the same three steps. First, the material must be put into its liquid or liquid crystalline state. This can be done by either heating the pure polymer or adding solvent to the pure polymer. Second, this fluid must be made to flow in a way that causes the polymer chains to orient and to produce the proper final shape. Third, the fluid must solidify, either by cooling the pure polymer or removing solvent from the polymer solution.

The second step is the most critical of the three. A typical way of heating a polymer and getting it to flow is to use a *screw extruder*. As shown in figure 9.13, a polymer in its solid form (usually pellets) is introduced into one end of the screw extruder and pushed along by the action of the screw. Heaters along the outside of the extruder heat the polymer as it moves, causing it to melt and mix thoroughly. At the other end is a *die*, which forces the molten polymer through narrow channels. The narrowness of the openings causes the pressure inside the extruder to rise as the screw turns. This combination of

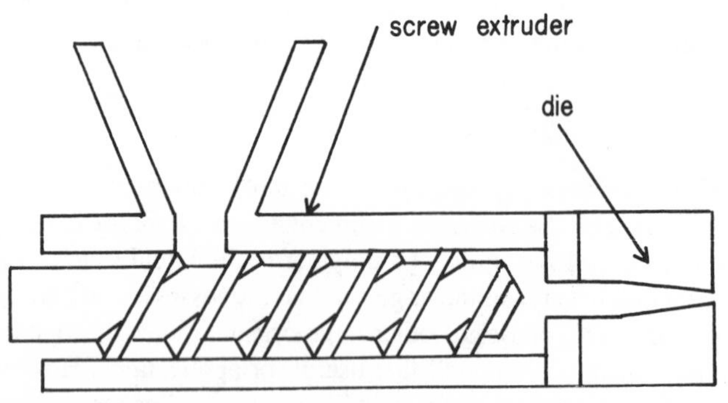

Fig. 9.13 Diagram of a simple screw extruder. The solid is introduced through the funnel, heated and mixed as it is moved along by the screw, and forced through the die.

high temperature, high pressure, and the geometry of the die is what causes the polymer molecules to orient. The highly-ordered fluid polymer exits the die (usually in the form of a thin sheet or fiber), where it cools quickly into a highly-ordered solid.

There are several variants to this procedure. In a process called *blow extrusion*, a jet of air directed against the film exiting from the die causes it to form a long cylindrical bubble before it solidifies. The polymer material can be made quite thin by this process. All of our plastic bags are made in this way. *Blow molding* is slightly different. A cylindrical piece of polymer is extruded around a thin tube that sits inside a mold. Compressed air is blown out of the thin tube, expanding the material until it takes the shape of the mold. Plastic bottles are made this way. If the fluid polymer is forced into a cavity of a certain shape to solidify, a solid object results. This process is called *injection molding*. Thin sheets of polymer can be fabricated by forcing the material to flow between two rollers. This *calendering* process can be used with bulk polymer to make thin sheets or with already extruded sheets to make them thinner. Finally, a process called *fiber spinning* produces threads made from twisted polymer fibers. Small orifices on a *spinneret* form multiple fibers that are twisted as they are drawn thinner and collected.

The ability of flow to orient polymer molecules results from the fact that the forces between the molecules are such that the molecules tend to align. In the absence of flow, the random motion of the molecules due to temperature overcomes these forces in the case of a liquid polymer and allows only a small amount of order in the case of a polymer liquid crystal. Flow induces additional orientation order that can overcome the random thermal motion in liquid polymers or increase the degree of orientational order in polymer liquid crystals. Of course, the conditions must be right for flow to have the desired effect. Before these processes can be used, therefore, much experimentation with different temperatures, polymers, flow speeds, die shapes, etc., must be done.

Polymer liquid crystals can achieve extremely high orientational order, which is useful for certain applications. The best example is Kevlar, which is a fiber formed from the liquid crystal phase of a polyamide. Although only slightly denser than nylon (which is formed from the liquid phase), Kevlar is roughly thirty times stronger. In fact, pound for pound (steel is five times denser) Kevlar is stronger than steel. No wonder it has found widespread use. Automobile tires, bullet-proof vests, and mooring cables are only a few examples of Kevlar's practical applications.

Polymer liquid crystals respond to electric and magnetic fields, which means they can be used for display purposes. An electric or magnetic field can orient a polymer liquid crystal, thereby changing its optical characteristics. If the polymer liquid crystal is cooled into the solid phase, it is possible to freeze in this order. Since in this case the order remains even after the field has been removed, a display using a polymer liquid crystal can have storage capability. For example, a laser can be used to produce local heating and raise the temperature into the liquid crystal phase. If no electric field is applied to the cell, the polymer cools in an unoriented state where the laser beam has been applied and thereby creates a permanent image. The image can be erased by applying the laser beam again, but allowing the polymer to cool as an electric field is applied to the cell. Polymer liquid crystals have some properties that make them very attractive for display purposes. Liquid crystal polymers are relatively inexpensive, stable, and most important, can be easily fabricated into thin films.

As mentioned before, liquid crystals are a useful medium in which to perform a separation of substances (chromatography). Liquid crystal polymers offer one advantage for such purposes, in that they release fewer molecules as gas when they are heated to high temperatures. Under these circumstances, liquid crystal polymers are much more efficient than other materials in separating substances.

Ultra-high-strength fibers and electro-optic displays are two very important areas of modern technology. Clearly polymer liquid crystals are destined to play an even larger role in these two areas in the future. However, there may be other significant applications around the corner. For instance, liquid crystal ordering of biologically important macromolecules is critical to the workings of cells. Is the function of certain structures within the cell, perhaps of RNA or DNA itself, affected by whether the macromolecules possess orientational order or not? The answer to this question seems to be yes, so knowledge of the processes involved might open the door to significant medical breakthroughs.

CHAPTER 10

Theories, Defects, and Fluid Lattices

Previous chapters have certainly demonstrated that liquid crystals have been used for important technical applications over the last thirty years. What has not been discussed is that the study of liquid crystals has been just as important scientifically. As scientists have struggled to understand why certain molecules form liquid crystal phases and have certain properties, they have found themselves asking some of the same questions as scientists studying other phases of matter. In fact, some important new ideas applicable to many phases of matter have been first shown to be valid in liquid crystal systems. In this chapter, I will first discuss some of the theories of both the liquid crystal phase itself and the transitions that take place between liquid crystal phases. I will then examine the defects that occur in liquid crystals and the reasons why they are important to our understanding of the liquid crystal phase. Finally, we will see how nature orders these defects to produce a truly unique phase of matter. In short, this is a chapter concerning the science of liquid crystals. Like all areas of science, the story of liquid crystal science is full of creative new ideas, brilliant experiments, and exciting surprises.

Theories of the Liquid Crystal Phase

An observer of nature recognizes quickly that just about every process is quite complex. Yet the success of science rests in the ability of the human mind to uncover fairly simple ideas, that allow us to see through this complexity and bring some order to diverse natural processes. In this sense, the achievements of science stand as a monument to the creativity, ingenuity, and perseverance of the human spirit. The attempt to understand the liquid crystal phase is an excellent example of this struggle to understand complex phenomena, because the liquid crystal phase results from the mutual interaction of an enormous number of molecules. Even with our fastest computers, calculations involving such a large number of molecules are not possible. Therefore, scientists must dream up ways to capture the behavior of this large number of molecules without considering each one of them. What follows is a description of a few of these attempts.

Perhaps the most influential theoretical attempt to describe the liquid crystal phase is due to two German scientists, W. Maier and A. Saupe. Although their theory was formulated in 1960, it continues to be the starting point for a large amount of theoretical work. The *Maier-Saupe theory*, as it is called, begins with the assumption that the most important force between liquid crystal molecules is the *dispersion* force. This force occurs between two molecules that possess no permanent electric dipoles, but that can possess induced electric dipoles. If the electronic structure of one molecule fluctuates (due to temperature effects) and gains a momentary electric dipole, the electric field of this dipole polarizes the other molecule slightly, thereby creating an induced electric dipole on the second molecule. This induced dipole on the second molecule produces an electric field that tends to strengthen the original dipole on the first molecule. The result of this interaction is that the initial spontaneous fluctuation results in two induced dipoles, which tend to (1) attract each other and (2) keep the two molecules aligned. Because the origin of this force is in the random fluctuations occurring in the electronic structure of molecules, the dispersion force is not exceedingly strong. For example, Maier and Saupe performed calculations indicating that the dispersion force is proportional to the inverse sixth power of the separation. Compared to the force between two charges that varies as the inverse second power, this decrease with separation is very rapid.

Even armed with the force between molecules, the Maier-Saupe theory must still deal with the problem of calculating what happens if millions of molecules mutually attract each other with this force. To overcome this problem, the theory takes the giant step of conjecturing that a single molecule in a sea of other molecules must experience a force that on average is the same for all molecules. That is, with so many molecules diffusing around randomly, the average effect on any one molecule must be the same for all molecules. The theory thus deals with the force a single molecule experiences in the sea of other molecules; and since it is the same for all molecules, one calculation will suffice. The force a single molecule experiences is obtained by combining conjecture with the fact that the basic force between molecules is the dispersion force. The conjecture is that a molecule will experience a strong force to point along the preferred direction if the molecules around it are highly oriented along this direction. If so, the force on a single molecule should be proportional to the order parameter S. This results in a force on a single molecule that represents an average or mean due to all the other molecules. For this reason the Maier-Saupe theory belongs to a class of theories called *mean field theories*.

To calculate how the order parameter changes with temperature, the Maier-Saupe theory calculates the thermodynamic average of $(3\cos^2\theta - 1)/2$ using the force on one molecule due to all the others. The average cannot be directly computed, because this force contains the order parameter S, which is the unknown that the theory is attempting to calculate. The average of $(3\cos^2\theta - 1)/2$ is just the order parameter S, so the equation for this average is really an equation that says S equals some expression that also contains S (along with temperature, volume, etc.). This equation can be solved for S in terms of temperature and the other parameters. In this way, the Maier-Saupe theory produces a prediction for how the order parameter should vary with temperature in the nematic phase. Because all the details of the molecules are lost when working with this simple average force on a single molecule, the Maier-Saupe theory produces a single prediction for all liquid crystals. For some liquid crystals the Maier-Saupe prediction corresponds well with measurements of the order parameter. For other liquid crystals, however, the agreement is not very good. An example of order parameter measurements in two liquid crystals together with a graph showing the prediction of the Maier-Saupe theory is shown in figure 10.1.

There have been many embellishments to the Maier-Saupe theory since 1960. For example, the average force experienced by a single molecule due to the other molecules has been derived using more forces than just the dispersion force. In another example, the fact that the molecules are not perfect rods but more like flattened rods has been taken into account in deriving the average force on a single molecule. The predictions of these theories depend on the relative

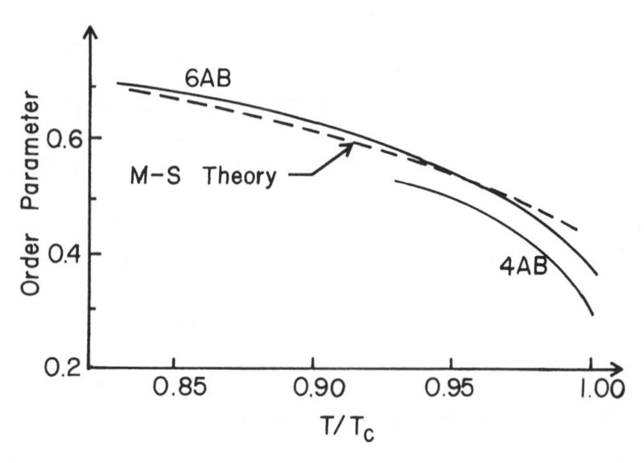

Fig. 10.1 Theoretical and experimental order parameter curves. The dashed curve is the prediction of the Maier-Saupe theory. The two solid curves show the result of order parameter measurements using two members of the PAA series. The horizontal axis is the temperature as a fraction of the nematic-isotropic transition temperature, which is different for the two compounds.

strength of the various forces or the relative dimensions of the molecules. By making the proper selection for these parameters in the theory, good agreement between theory and just about all order parameter measurements has been obtained. The only problem is that it is difficult to select these parameters only from knowledge of the molecule being investigated and thereby predict how the order parameter should vary for a particular liquid crystal system.

Another theoretical approach that yields successful results emphasizes repulsive forces between molecules rather than attractive forces. When two molecules get very close together, the electrons of one molecule are close to the electrons of the other molecule, and the repulsive forces between these electrons dominate any other forces between the two molecules. Since these forces are relatively weak when the separation is not very small, but increase sharply at some very small separation, these forces in effect keep the two molecules from occupying the same space. For this reason, these forces are called *steric* forces. Theories starting with steric forces seek to describe a system of rods that cannot penetrate each other. Such systems tend to align the long axis of the molecules, because the more they are aligned, the less chance there is for two molecules to collide.

The problem of performing the calculation for a huge number of molecules must also be overcome in this approach. Instead of performing averages over all molecules to calculate, for instance, the energy of the system, some theories consider only pairs of molecules. Other theories examine the situation in cases where the average is not difficult to perform, i.e., for molecules much smaller or much larger than all the other molecules. An intermediate result that gives the correct behavior in the case of both large and small molecules is then selected for use in the rest of the calculation. One important aspect of these theories stems from the fact that the forces depend only on the shape of the molecules and the average distance between them. Temperature does not come into the calculation at all. Therefore these theories are not appropriate in describing thermotropic liquid crystals, but work well in situations where temperature is not so important. For example, these theories give results that compare favorably with measurements on a solution of rod-shaped macromolecules in a solvent. In this case the concentration of macromolecules is the important quantity, which can easily be taken into account by the theory. Both theory and experiment agree that at high enough concentrations of macromolecules, the molecules prefer an aligned liquid crystal phase over an unaligned isotropic phase.

One related type of theory is the *lattice model*. In these theories, the space the liquid crystal occupies is considered to be full of lattice points. A "molecule" therefore occupies a line of lattice points. The theoretical scientist considers all the different ways liquid crystal "molecules" can be placed in this space with each occupying a line of lattice points, but with no two "molecules" penetrating each other. An example of such "molecules" in a two-dimensional lattice is shown in figure 10.2. Averages over all the different possibilities are performed (normally requiring a computer), followed by the calculation of various properties of the system. Since only steric forces are contained in this theory, the results are most appropriate to lyotropic liquid crystalline systems. It should also be pointed out that the size of the lattice cannot be very large, before the computer calculation begins to take too long a time. A cubic space with roughly one hundred lattice points on a side is about the present limit.

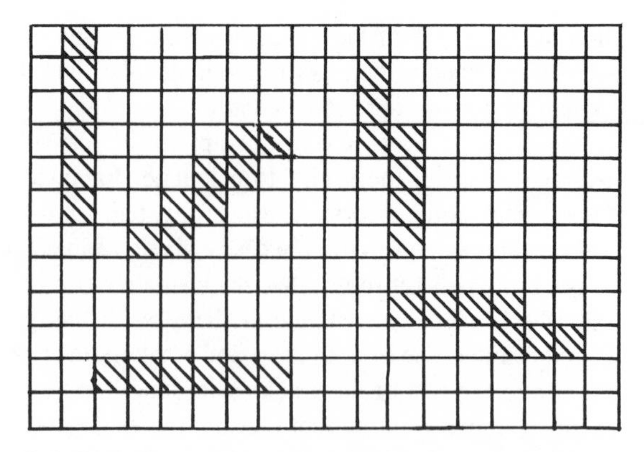

Fig. 10.2 Several examples of how molecules with various orientations can be represented in a lattice model.

Another type of theory utilizes the *Monte Carlo* method. This is a method that computes the averages by artificially generating all the possibilities, but in a way that maintains consistency between the distributions of all the relevant parameters. How this is done is quite clever, and the method can be illustrated with a simple example. Suppose you want to find the value of $\pi(3.1416)$ using the Monte Carlo method. This can be done by considering the square and the unit circle shown in figure 10.3. The area of the square is 4 and the area of the unit circle is π (area equals π times the radius squared). If you generate dots that are spread out throughout the square uniformly, the ratio of the areas ($\pi/4$) will give the fraction of the dots that fall within the unit circle. So on a computer you generate pairs of numbers (each randomly between -1 and 1) and call the first the x-coordinate and the second the y-coordinate. Eventually these numbers will fill the unit square uniformly. For each pair, you check if the point falls within the unit circle (is $x^2 + y^2$ less than 1?), and after many trials compute the fraction of points that fall within the circle. If you multiply this number by 4, you'll get a value for π. How good this value is depends on (1) the number of trials you perform, and (2) the quality of your generator of random numbers. Note that the points in the square are generated without considering anything about the situation, but the probability of a point falling inside the circle is the key in using the generated numbers to get the desired quantity.

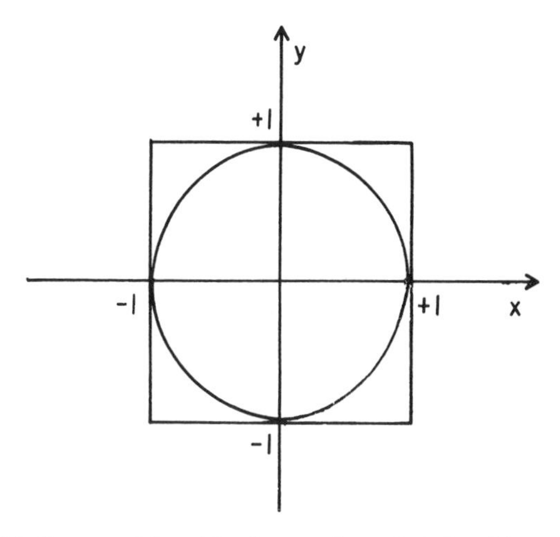

Fig. 10.3 Square with a side of two units and circle with a radius of one unit. The value of π can be found from these two figures using the Monte Carlo method.

In liquid crystal theory, a point inside the square is replaced by a possible configuration of all the molecules in a hypothetical liquid crystal using a lattice model. Such a possible configuration is generated without any regard for whether it is likely or not. The value for the quantity being examined (for example, the order parameter) is calculated for this configuration but it is sometimes included in the average for all configurations and sometimes not, depending on whether it "wins a bet" against a generator of random numbers. The odds of "winning" (and therefore being counted) are set so that over enough trials the quantity is counted on average in accordance to the likelihood of that configuration actually being adopted by the molecules. If many configurations are generated (a million or so), the resulting average correctly reflects the likelihood of each of the possible configurations actually being adopted by the molecules, and a good average for the quantity is produced. Monte Carlo theories of the liquid crystal phase can include various forces between molecules and produce results that sometimes agree extremely well with experimental data.

It should be pointed out that although the fastest computers cannot calculate exactly what happens to a huge number of molecules when they mutually attract and/or repel each other, we are getting a bit closer. It is now possible for the "exact" motion of each of a "not too large" number of molecules to be followed, taking into account all mutual forces. The configuration of molecules is actually calculated for small steps in time; these configurations are averaged to produce whatever quantity is desired. The chief limitation of this *molecular dynamics* method is that the number of molecules is only around one thousand. Faster computers are sure to drive this number up and produce "exact" predictions starting with molecular forces.

Can the results of all this theoretical research be summarized? In fact, a few general statements are possible. First of all, indications are that longer range attractive forces and shorter range repelling forces are *both* important in thermotropic liquid crystals. However, there are reasons to suspect that attractive forces in smectic liquid crystals are more important than in nematic or chiral nematic liquid crystals. The situation is quite different in lyotropic liquid crystals and polymer liquid crystals in solution, where steric forces seem to be by far the most important. Thus once more we see that the liquid crystal phase literally results from a delicate balance between opposing forces. The delicateness of this balance is beautifully displayed in certain experiments done under pressure. As mentioned in the beginning of chapter 8, increases in pressure more or less act like decreases in temperature, as far as the stability of certain phases is concerned. Usually increasing the pressure in the isotropic phase causes a phase transition to the nematic phase and then to the smectic phase. However, in some liquid crystals the sequence upon increasing pressure is isotropic, nematic, smectic, and nematic again. This second nematic phase is called the *reentrant nematic* phase and results because the increase in pressure decreases the intermolecular separation, thus increasing the repelling steric forces. In certain cases the molecular geometry produces an increase in steric forces large enough to throw the balance back in favor of the nematic phase.

Liquid Crystal Phase Transitions

One of the most exciting areas of both theoretical and experimental physics over the last twenty-five years has been the study of phase transitions. Although we have been talking about phase transitions in which physical properties such as density and index of refraction change, there are many other types of phase transitions in which other physical properties change. For example, the magnetic properties of a solid can change drastically at a certain temperature. Below this temperature the solid can be made into a permanent magnet, whereas this is not possible at a higher temperature. Likewise, the magnetic and electrical properties of a material change at the phase transition between a normal conductor and a superconductor. What the recent research has told us is that many of these phase transitions have much in common. In fact, some of them are virtually identical, even though the physical properties involved and the underlying mechanisms causing the phase transition are completely different. These results seem to imply that the fact that all these transitions involve large numbers of interacting elements makes many details of the system unimportant. Instead, only a few characteristics of a phase transition determine its nature. There are many different possible liquid crystal phase transitions, so for this reason they have played a significant role in the study of phase transitions in general.

One of the characteristics of a phase transition that is important is the *symmetry* of the phases on each side of the phase transition. Symmetry is an extremely important concept in science and is worth discussing. By symmetry, scientists are referring to what operations can be performed on an object without it changing in any way. For example, consider the objects in figure 10.4. The rectangle shown in figure 10.4(a) can be rotated through 180° about any of the three axes and it will appear exactly as before. Reflection is another operation. Reflection through a plane means all points on one side of the plane are moved perpendicularly through the plane to a point equidistant away on the other side of the plane. Reflection of the rectangle through the two planes shown also leaves the rectangle unaltered. Note that a square has four axes where 180° rotations leave it the same, but it also has one axis where a 90° rotation brings the square back on itself. The square also has four reflection planes instead of two, which have been omitted from figure 10.4(b). Obviously the square has more symmetry than a rectangle. Figure 10.4(c) shows a piece of a rectangular lattice. All the symmetry of the rectangle is present in the rectangular lattice, because the lattice is made up of

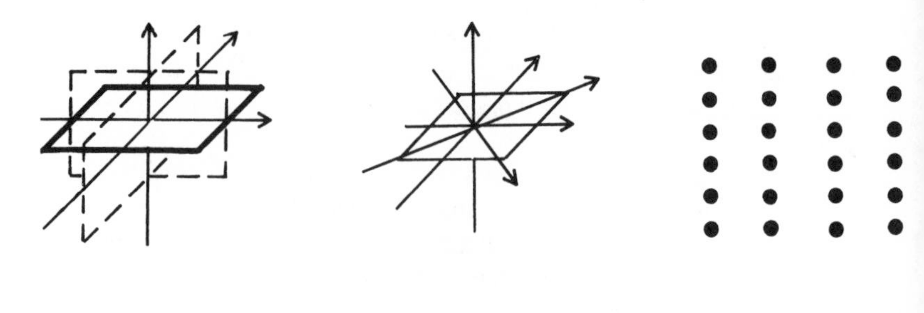

Fig. 10.4 Symmetry operations of a rectangle and square. Three 180° rotation symmetry axes and two reflection symmetry planes are shown for the rectangle in (a). Four 180° and one 90° rotation symmetry axes are shown for the square in (b). The four reflection symmetry planes are not shown. A rectangular lattice is shown in (c). Two translation symmetry operations are present in the lattice in addition to the symmetry operations shown in (a).

rectangles. But notice that the entire lattice can be translated in certain directions by the right amount and remain unchanged. These translations are therefore also *symmetry operations* for this lattice.

The symmetry of a phase is determined by how that phase can be translated, rotated, reflected, etc., and still look exactly the same. For example, the isotropic phase has continuous rotational symmetry about every possible axis, in that you can rotate a piece of isotropic liquid about any axis and the phase looks exactly as it did before the rotation. The nematic phase is quite different. There is one axis, the director, about which you can rotate the phase through any angle and not change the phase at all. However, rotation about any other axis usually results in a nematic sample with a director pointing in a different direction. Because the phase is no longer identical after the rotation, we say that continuous rotational symmetry does not exist about that axis. When an isotropic liquid changes to a nematic liquid crystal, some of the symmetry present in the higher temperature phase is lost in the lower temperature phase. This is a general feature of all phase transitions and is referred to as *spontaneously broken symmetry*.

The nematic to smectic transition is another good example of the importance of symmetry. In the nematic phase, there is complete translational symmetry, in that moving the sample through space (without rotation) in any direction does not change the properties at any point fixed in space. This is not true in the smectic phase. Translation parallel to the layer planes keeps the properties at all points the same, but translations perpendicular to the layers can cause properties at certain points in space to change. A symmetry present in the nematic phase is therefore lost in the transition to the smectic phase.

One other important characteristic of a phase transition is the nature of the order parameter involved. Although we have been using the term order parameter to describe the orientational order in liquid crystals, it is really a much more general term. An order parameter can be described for every phase transition. In the higher temperature phase, the order parameter is zero; in the lower temperature phase, the order parameter is nonzero. For example, the order parameter appropriate for the magnetic phase transition is the magnetization (magnetic dipole per unit volume). Above a certain temperature the material has no net magnetization so the order parameter is zero. At the phase transition temperature, the material becomes magnetic with a nonzero magnetization (nonzero order parameter). In a similar vein, the order parameter appropriate for the nematic-smectic *A* phase transition must be different from the nematic order parameter S, because S is nonzero both above and below the nematic-smectic *A* phase transition. The appropriate order parameter must describe the fact that the molecular centers are not uniformly distributed in the smectic phase, but instead are bunched into layers. An order parameter that describes a density wave (high density of molecular centers alternating with low density of molecular centers) does the trick. In some transitions (called discontinuous or *first order* phase transitions), the order parameter changes from zero to a finite value at the transition. In other transitions (called continuous or *second order* phase transitions), the order parameter increases from zero gradually at the transition. Some liquid crystal transitions are first order (isotropic-nematic) while others are sometimes second order (smectic *A*-smectic *C*).

This discussion reveals why liquid crystals are important for the study of phase transitions in general. The different liquid crystal transitions afford many possible types of transitions, all with different order parameters. Modern research has revealed that the symmetry of the phases and the mathematical nature of the order parameter are just about all that determine what happens at the phase transition, regardless of what system is under consideration or what properties change at the transition.

Defects in Liquid Crystals

A sample of a nematic liquid crystal normally does not possess a director that points in the same direction at all points in the sample. In a certain area of the sample it might point in one direction, and at another it might point in a different direction. Often there is a place between these two areas where the director abruptly changes the direction in which it points. It is impossible to define the direction the director is pointing at the location of abrupt change, so this represents a *defect* in the order of the liquid crystal. Point defects and line defects are the most important types of defects in liquid crystals. As we shall see, the defects present in a liquid crystal sample viewed between crossed polarizers in a microscope determine more than any other factor the appearance of the liquid crystal.

It is important to realize that an abrupt change in the direction of preferred orientation always implies severe distortion of the director configuration in the vicinity of the defect. Previously I discussed how forces are necessary to distort the director configuration, with greater distortions requiring greater forces. These forces would be released if the defect were removed and replaced with a constant director in the region. The fact that defects are stable indicates that outside influences on the director configuration (electric field, nearby glass surface, etc.) are not compatible with a constant director in the region, so a director configuration with a defect is the best compromise possible.

Point defects are less common than line defects. They sometimes do occur in thin capillary tubes and spherical droplets. The most simple example is a spherical droplet in which the molecules are constrained to be perpendicular to the surface of the droplet. In the center a point defect might occur. As we saw in chapter 6, if the molecules lie parallel to the surface of the droplet, two point defects occur (see figure 6.13). An interesting example occurs in a thin capillary tube where the molecules must orient themselves perpendicular to the cylindrical surface of the tube. A diagram of a capillary tube is shown in figure 10.5(a), where two point defects are depicted. What is interesting about the two point defects is that the director configurations above the top one and below the bottom one are the same. That means that it must be possible for the two point defects to combine and annihilate each other, resulting in the director configuration (with no defects) shown in figure 10.5(b). The attraction and subsequent annihilation of these two defects is analogous to what happens when two opposite electric point charges attract and "cancel" each other. For this reason, scientists sometimes assign one of these defects a plus sign and the other a minus sign.

One other possibility for the capillary tube with the director perpendicular to the surface is that a line defect may be present running along the center of the tube. This is shown in figure 10.5(c). Line defects were given a special name by the early liquid crystal researchers—*disclinations*. The name stems from the fact that the line repre-

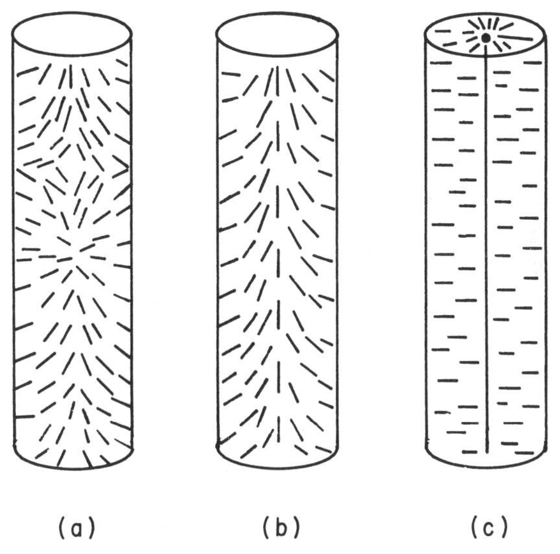

Fig. 10.5 Defects in a capillary tube. Two opposite point defects are shown in (a); the director configuration in (b) has no defects; and a line defect is shown in (c).

sents a "discontinuity" in the "inclination" of the director. Many different disclinations are possible. To classify them, we simply examine the director configuration in the plane perpendicular to the direction of the line. Figure 10.6 contains six examples. Notice that signs have been assigned to indicate which disclinations are opposite. The numbers represent the *strength* of the disclination. By comparing figure 10.5(c) to the examples in figure 10.6, you should be able to conclude that the line defect in the capillary tube is in fact a disclination of strength +1.

One important feature of disclinations is obvious from figures 10.5(b) and 10.5(c). A disclination of strength +1 can always relax to the structure of figure 10.5(b) where no defect is present at all. In this sense, a +1 disclination is not absolutely stable. This is true for all disclinations of integer strength, whether positive or negative.

As discussed in chapter 4, places where the director is parallel to the crossed polarizer or analyzer axes appear dark under the microscope. Since defects represent regions where the director configuration changes, these dark regions must also have a unique appearance around defects. To imagine what these dark regions would look like

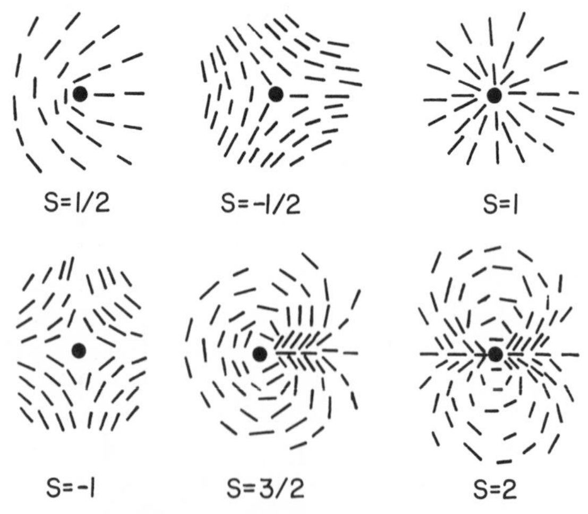

Fig. 10.6 Six different types of disclinations. The numbers give the strength of the disclination. Disclinations of opposite sign can combine to give a disclination of lower strength.

for each of the disclinations shown in figure 6, locate the regions around each disclination where the director points either horizontally or vertically. This has been done in figure 4.11 for a disclination of strength +1. Notice that each of the disclinations in figure 10.6 possesses a pattern of dark regions that is unique, allowing for easy identification. A number of the plates contain such characteristic dark areas. Can you recognize them and identify the strength of the disclination?

Defects in chiral nematic and smectic liquid crystals differ from those in nematics due to either the twisted or layered structure. Chiral nematics and smectics can be deformed in many ways, but the most difficult deformation is one that changes the pitch of a chiral nematic or the layer spacing of a smectic. It is therefore not surprising to find that the defects that do occur in these two types of liquid crystals are the ones that do not involve either of these deformations.

Figure 10.7 illustrates some of the types of disclinations that occur in chiral nematic liquid crystals. The disclination shown in figure 10.7(a) is an important one, since an additional half turn of the chiral nematic helix is introduced at the point of the disclination. This type of disclination shows up in wedge-shaped samples of chiral nematic liquid crystals with the helical axis perpendicular to the surfaces. As the space between the two glass surfaces increases, more turns of the helix fit in the space between the glass. To keep fitting in more turns

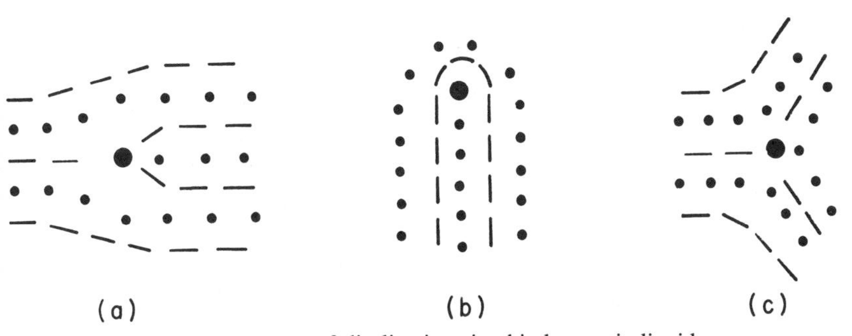

Fig. 10.7 Three types of disclinations in chiral nematic liquid crystals. The dots represent regions where the director points into or out of the page. The lines represent regions where the director lies in the plane of the page. The disclination is shown as a large dot.

of the helix, disclinations must appear at regular intervals in the sample. As discussed in chapter 5, these disclinations are quite visible under the microscope and are called Cano lines. The circular lines in plate 8 are therefore disclinations of the type shown in figure 10.7(a).

Under certain conditions, chiral nematic and smectic liquid crystals form a very interesting texture when placed between two pieces of glass. In both cases, disclinations are responsible for the appearance of this texture. Let us consider smectic liquid crystals in the following discussion, but keep in mind the same description holds for chiral nematic liquid crystals. The only difference is that the layer spacing in the case of smectics is analogous to the pitch in chiral nematics.

Imagine regions of smectic liquid crystal growing in a nematic because the temperature is decreasing. The smectic may be distorted as it grows, but in ways that keep the distance between the layers constant. A typical director configuration is shown in figure 10.8(a). Notice that there are two disclinations running perpendicular to the page, with a disclination in the plane of the page between them. Often the two disclinations running perpendicular to the page are actually part of the same disclination that bends around in an ellipse. This is shown in figure 10.8(b), where another view of the texture is presented. The third disclination running through the ellipse has the shape of a hyperbola. The presence of these conic sections (plane slices through a cone) is the reason this texture is called the *focal conic texture*.

Fluid Lattices

When Reinitzer described the melting behavior of cholesteryl benzoate in 1888, he stated that the substance briefly turned blue as it changed from clear to cloudy. Although researchers working with chiral nematic liquid crystals must have noticed this phenomenon, extremely little appeared in scientific literature for over eighty years. A few reports of experiments were published during the late sixties and early seventies, but the situation changed dramatically when detailed experiments revealed that the blue color was due to at least two new liquid crystal phases. Suddenly these *blue phases* became the source of widespread interest, and experiments soon showed that the structures of these phases were very different from all other liquid

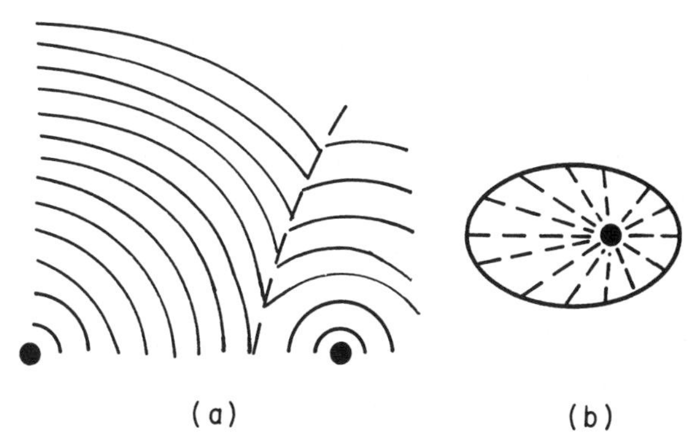

Fig. 10.8 Origin of the focal conic texture in a smectic liquid crystal. (a) shows smectic layers growing from two points on a glass surface. These points are actually part of a disclination in the shape of an ellipse. (b) shows the view from above the glass surface, where this elliptical disclination is drawn as a line. The disclination shown as a dashed line in (a) appears as a dot in (b). Plate 4 is a picture of a texture composed of these structures.

crystal phases. New theoretical ideas were born and entire classes of new experiments were performed on liquid crystals. As we shall see, defects do not just occur at various points in these blue phases; they are in fact essential to its structure.

Before scientists made progress in understanding the blue phases, they had to realize that an old assumption was not always valid. I have already discussed that the helical structure of chiral nematic liquid crystals (figure 1.6) results from the fact that the molecules prefer to be at a slight angle to one another, rather than parallel. For almost one hundred years, scientists assumed that this structure (that involves a single helical axis about which the director rotates) is the most stable structure for such molecules. It turns out that another structure can be more stable. Instead of a single helical axis, the director in this alternative structure rotates in helical fashion about every axis perpendicular to a line. Such a structure has been given the name *double twist*, although an infinite number of helical axes are really present. Figure 10.9 shows the double twist structure. The plane containing the helical axes is the plane of the page in figure 10.9(a). The director points into or out of the page in the center, and

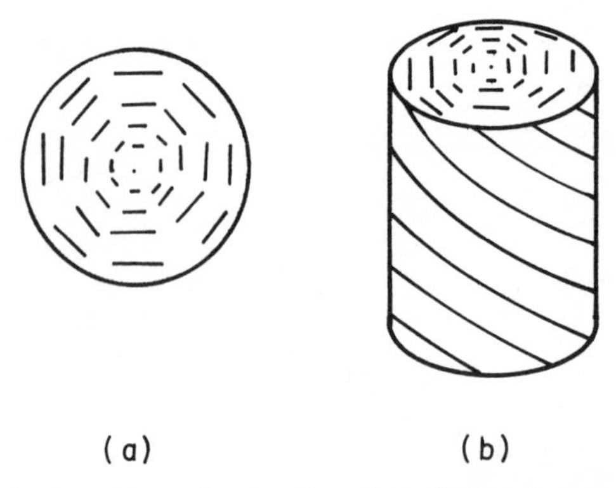

Fig. 10.9 Two views of a double twist cylinder. A cross-sectional view (a) shows that in the center the director is parallel to the cylinder axis. The director twists in going away from the center in any direction. As shown in (b), the director has twisted by 45° between the center and outside of the cylinder.

rotates as you move outward from the center in any direction. There is no change in the director going into or out of the page. A perspective drawing is shown in figure 10.9(b). The curved lines on the outside of the object represent the fact that the director has rotated 45° in going from the center to the outside.

There is a good reason why this structure does not occur often even though it is more stable than the normal helical structure or *single twist* structure. The double twist structure is more stable only within a small distance from the line at the center. This means that a large double twist structure is not more stable than a large single twist structure. Since this small distance is on the order of the pitch of the chiral nematic liquid crystal (roughly 0.0005 mm), and since most samples of liquid crystal are much larger than this, it is not surprising that the double twist structure rarely occurs.

The blue phases are special cases where the double twist structure does fill up a large volume. If you end the double twist structure in all directions from the central line at points where the director has twisted by 45°, a *double twist cylinder* results. This is the structure that appears in figure 10.9(b). Because its radius is small, this cylinder is more stable than an equal volume of single twist chiral nematic. A large structure can be made from these double twist cylinders, but the problem is that defects occur at regular points between the cylinders. These defects tend to make the structure less stable, but in one of nature's most delicate compromises, the structure composed of double twist cylinders and defects is slightly more stable than the single twist structure without defects, but only for a temperature interval roughly 1°C wide just below the transition from the chiral nematic phase to the isotropic liquid. Two examples of structures using double twist cylinders are illustrated in figure 10.10.

Notice that these structures possess a regular array of defects at places where double twist cylinders with different directors meet. In the structures of figure 10.10, the defects have a cubic arrangement. In figure 10.10(a) the defects are positioned at the corners of stacked cubes; in figure 10.10(b) the defects occur at the corners and the middle of stacked cubes. These defects are therefore arranged in a lattice, just as the molecules in a crystal are. There is a big difference between a crystal and this lattice of defects, however. In a crystal, a physical entity (a molecule) occupies each lattice point. In these structures, no physical entity, merely a defect in the director configuration, sits at each lattice point. The blue phases are these fluid lattices of defects.

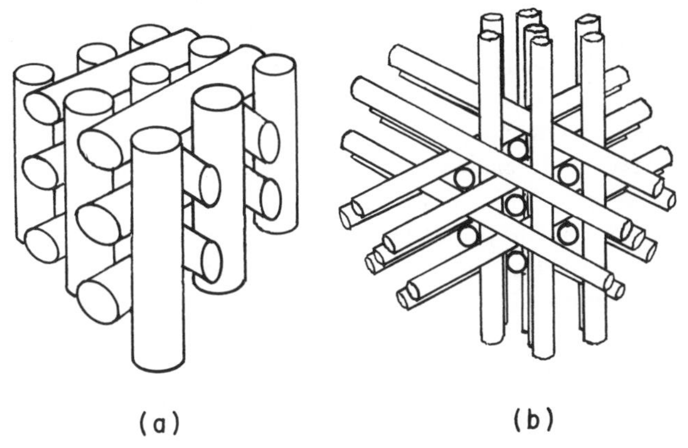

Fig. 10.10 Two structures composed of double twist cylinders. The orientation of the director at the point where two cylinders touch is the same for both cylinders. Defects are present only in the regions where three cylinders meet. These defects form a simple cubic lattice in (a) and a body centered lattice in (b).

The reason why crystals reflect X-rays can also explain why a blue phase is capable of reflecting a color (only some blue phases are blue): the spacings between the defects in a blue phase are roughly equal to the wavelength of visible light. For reflections of light off the various planes of defects, constructive interference occurs just as it does in crystals where the spacings of atoms are equal to the wavelength of X-rays (see chapter 5). A sample of one of the blue phases will usually have regions in which the defect lattices are oriented in different directions. When you observe these blue phases under a microscope using reflected light, these different regions reflect different colors because the wavelength for constructive interference depends on both the spacing between the lattice planes and the angle between the light beam and the lattice planes.

These fluid lattices have properties that are extremely similar to crystals. If a substance with a blue phase is cooled extremely slowly so that the blue phase lattice forms very slowly, the lattice will display its cubic arrangement by forming what looks like a single crystal. If the lattice planes are perpendicular to the viewing direction, the single crystal has the shape of a square. Plate 14 is a picture of single crystals of what is called the first blue phase. The cubic lattice of these crystals is turned through an angle relative to the viewing direction, so the single crystals do not have a square shape. In some cases even the facets of these single crystals are visible. Transitions from one blue phase to another force the single crystal to change its orientation and lattice spacing. This can cause deformation and regular defects in the single crystal (not to be confused with the defects that make up the lattice), as is evident in the crystal of plate 15. The crystal shown in this picture was originally square and uniform, as is common for what is called the second blue phase. However, lowering the temperature caused a transition to the first blue phase, which produced the intricate pattern of defects evident in the photograph. These pictures certainly are reminiscent of how solid single crystals behave. It is difficult to look at these pictures and remember that the molecules are not locked into place. Molecules diffuse in and out of the single crystal, but the orientational order of a molecule changes abruptly when it passes from one phase (outside the single crystal) to the other (inside the liquid crystal).

A new technique for viewing these single crystals has recently been used, and it provides much more information about the structure of the various blue phases. In this technique, the lenses of the microscope are used in a slightly different way. Normally, the objective lens of a microscope is placed close to the sample so that a magnified image forms on the other side of the objective lens. An eyepiece lens then focuses on this image, magnifying it further for viewing. In this new technique, the image that is magnified for viewing is not the magnified image of the objective lens, but the light at a point behind the objective lens called the focal plane. If the light entering the sample consists of only one wavelength and is not convergent, then the only way light can get to any part of this focal plane (other than the center) is if constructive interference by lattice planes in the sample causes light to exit the sample at a different angle. This one-to-one correspondence of light at a point in the focal plane to a set of lattice planes in the sample is what makes this technique so powerful.

A picture of the light in the focal plane usually consists of a set of lines. These lines are called *Kossel lines*, after a German physicist who used a similar procedure in X-ray experiments around 1925. An example of the Kossel lines from a blue phase single crystal is given in plate 16. Notice how the picture reveals how the lattice planes are arranged in a very symmetrical way. Likewise, different lattice planes can be examined by using light of different wavelengths. This technique has been used extremely effectively in studying how the presence of an electric field distorts the lattice in a blue phase single crystal.

As if two blue phases with two different types of fluid lattices are not enough, nature seems to have come up with a third possible way to find a delicate balance between the double twist structure and defects. A third blue phase exists, yet it does not seem to have the same type of cubic lattice as the other two blue phases. There are experimental results which demonstrate that this third blue phase has some of the properties of the other two blue phases, but there is other experimental evidence which indicates that this third blue phase has no regular lattice. Perhaps it is a fluid lattice analog of an amorphous solid with a fairly random array of lattice points. The structure of this third blue phase remains a perplexing problem for today's scientists because it seems to have less order than any other liquid crystal phase, yet there is good evidence that it is as different from the isotropic liquid phase as many other liquid crystal phases. Researchers will eventually determine the structure of the third blue phase, but it will probably take some new theoretical ideas and some clever experiments.

CHAPTER 11

Biological Importance of Liquid Crystals

Most of us have heard the statement that biological organisms are 90% water. After hearing such a statement, how many of us then ask how organisms can produce structures that have some degree of rigidity? After all, water has no rigidity at all, yet fairly rigid structures are the norm in the biological world, not the exception. It is important to realize that I am using the word rigid in a very general sense. In addition to the normal meaning of ''stiff or unyielding,'' the word rigid when applied to a biological structure can also describe its ability to confine the movement of material or provide a substrate for the ordering of molecules. The ability of organisms to create rigid structures in an environment rich in water is part of both the development of single cells and the specialization of groups of cells. A single cell, a group of cells performing a specialized function, or an organ are all examples of biological structures, and understanding how such structures form is a central concern of biology. We shall see in this chapter that liquid crystals play a crucial role, not only in the formation of specialized structures, but also in the workings of every cell.

BIOLOGICAL STRUCTURES

All biological structures must form in an *aqueous* environment, i.e., an environment rich in water. This restriction is actually not that severe. For example, many substances can form crystalline structures in water if the concentration of the substance reaches the saturation point. Chemical reactions in water can produce a product insoluble in water, which then forms a precipitate. Under certain conditions the precipitate forms a rigid structure. Such processes are in fact biologically important, especially in the formation of solid structures, such as bone and shell. Yet these types of structures are not typical of the majority of biologically important structures for a very important reason. Solid structures do not allow molecules to diffuse about so that all the chemical and physical processes necessary for life can take place.

Thus the second requirement of most biological structures is that they be fluid. Only in a fluid do the molecules move from place to place in a way that allows for interactions with other molecules. Think of all the enzymes and proteins in your body. These molecules are produced in one structure, somehow move out of the structure and are transported to another part of your body, diffuse into another structure, and finally find the right place inside or near a certain cell. For this to occur fast enough, the environment must be highly fluid. This same process takes place on a smaller scale inside each cell. Substances go in and out of the various structures inside the cell. These structures (nucleus, mitochondria, endoplasmic reticulum, Golgi apparatus, etc.) work together to make the cell function properly just as the different organs of the body work together. Thus the ability of biological organisms to form fluid structures of some rigidity in an aqueous environment is central to the existence of life on this planet.

From what we have already discussed, you have probably already guessed that liquid crystals are a natural for such structures. Since the environment is rich in water, all of the structures discussed in chapter 8 should come to mind. Amphiphilic compounds spontaneously form highly ordered structures in a polar solvent such as water, and these structures are liquid crystalline (fluid) rather than solid. This means that there is constant diffusion of both the amphiphilic molecules throughout the structure, and water into and out of the structure. The rigidity of these structures varies over a wide range. Double layers of phospholipid molecules (figure 8.3[b]) are not very rigid at all, while twisted filaments of cylindrical micelles can be very rigid. Both serve extremely important biological needs.

In short, biological structures must be rigid enough to function properly and fluid enough to allow all the necessary processes to take place. There are not that many candidates for such structures. We shall see in this chapter that this delicate balance between rigidity and fluidity is achieved in part by liquid crystalline structures.

The Cell Membrane

The membrane on the outside of every animal cell serves two functions. It confines the contents of the cell to a certain limited volume and it controls the flow of ions and molecules into and out of the cell. Plant cells also possess a cell membrane, but an additional structure called the cell wall surrounds the cell membrane and provides rigidity in the normal sense of the word. In order for the cell membrane to confine the contents of the cell and maintain the shape of the cell, it must possess some rigidity. To control the transport of substances into and out of the cell, the cell membrane must also be somewhat ordered. A disordered structure cannot discriminate between various substances and therefore is incapable of controlling the movement of certain molecules. Finally, the cell membrane must control the movement of materials and the reactions taking place near the membrane quite rapidly. This can only be achieved if the molecules in the cell membrane can diffuse about quickly, which is only possible if the membrane is fluid. Liquid crystalline structures have these characteristics, and the fact that the cell membrane is a liquid crystalline structure is the most important implication of liquid crystals for the field of biology.

The standard model for the cell membrane is called the *fluid mosaic model*. As shown in figure 11.1, the primary structural component of the membrane is a bilayer of phospholipid molecules. Embedded more or less at random in the bilayer are proteins (long sequences of amino acids), which are responsible for the many reactions that occur in the vicinity of the cell membrane. For example, some proteins act as enzymes, while other proteins function as pumps, transporting material into and out of the cell. Some of these proteins lie on the inner or outer surface and some extend into and through the bilayer. These latter proteins must possess both hydrophobic and hydrophilic sections in order to span the bilayer. The two other main substances found in the cell membrane are cholesterol and water. The cholesterol is found mostly in the hydrophobic region between the polar head groups, whereas just about all of the water is found bound to the head groups. Plant cell membranes contain sitosterol and stig-

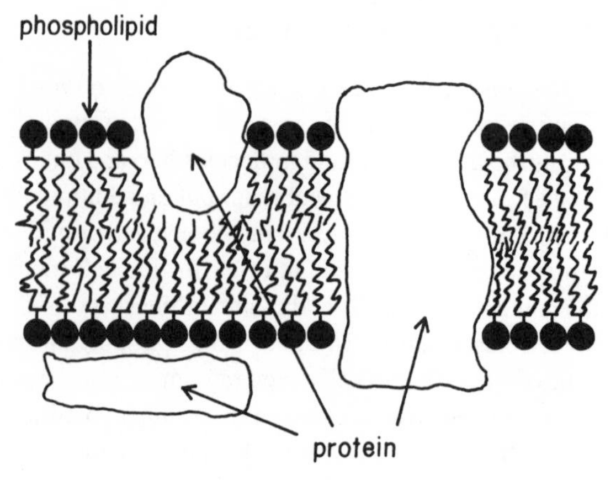

Fig. 11.1 Schematic diagram of a cell membrane, showing the phospholipid bilayer and embedded proteins.

masterol instead of cholesterol (different steroids) and glycolipids instead of phospholipids (different head groups).

The phospholipids are in the liquid crystal phase, meaning the head groups are not arranged in an array and the hydrocarbon chains are not rigid. The phospholipid molecules diffuse about the cell membrane, as do the proteins (albeit much more slowly). All this motion allows the reactions and interactions necessary for the proper functioning of the cell to take place. At the same time, the ordered structure provides a substrate in which the proteins can order in a specific way, which is also necessary for them to function.

Only certain types of phospholipids are found in animal cell membranes. *Diphosphatidylcholine* is the main ingredient of animal membranes, while *diphosphatidylethanolamine* predominates in the cell membranes of bacteria. As evident from figure 11.2, these two phospholipids have different head groups. In addition, the hydrocarbon chains usually have an even number of carbon atoms, with sixteen, eighteen, and twenty carbon chains accounting for 98% of the phospholipids in animal membranes. The majority of the bonds between the carbon atoms are single bonds where the carbon atoms share one pair of electrons. Double bonds, in which two electron pairs are shared, are sometimes present.

Ingenious experiments have demonstrated just how crucial the liq-

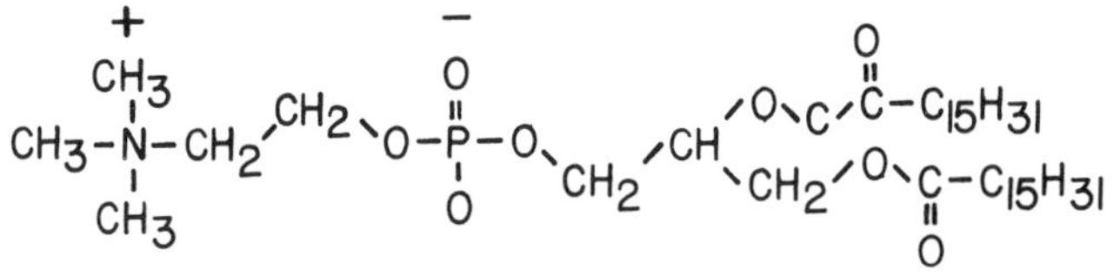

Fig. 11.2 Two typical phospholipids found in cell membranes: (a) diphosphatidylcholine and (b) diphosphatidylethanolamine.

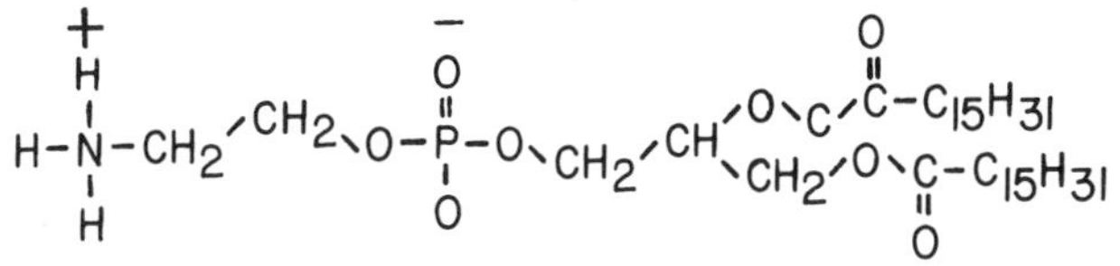

uid crystalline nature of the bilayer is to the proper functioning of cells. In chapter 8, I discussed how phospholipid bilayers possess a phase called the gel phase at temperatures below the liquid crystal phase. At the transition from the liquid crystal phase to the gel phase, the fluid hydrocarbon chains straighten and the polar head groups arrange themselves into a hexagonal array. The experiments consist of simply allowing a primitive organism to grow at a certain temperature and after a while extracting the membranes to examine at what temperature the liquid crystal to gel phase transition occurs. If the organism is forced to grow at a temperature of 35°C, the membranes have a transition temperature slightly below 35°C. If the organism grows in a 30°C environment, the membranes possess a transition temperature slightly below 30°C. Since the membranes have a transition temperature just below the ambient temperature, the membranes are all in the liquid crystal phase. Lowering the temperature below the transition temperature always changes the functioning of the cell membrane and often leads to the death of the organism. Work done with bacteria that live in the deep ocean trenches where the pressure is much greater than atmospheric pressure illustrates the need for liquid crystalline cell membranes. High pressure tends to raise the transition temperature. Organisms growing under high pressure therefore produce cell membranes that have transition temperatures just below the ambient temperature when under high pressure, but have transition temperatures far below the ambient temperature when the pressure is reduced to atmospheric values. Many of these organisms therefore die when brought to the surface of the ocean. This implies that it is also important that the gel to liquid crystal transition temperature not only be below the ambient temperature, but also be fairly close to the ambient temperature.

How can an organism control the transition temperature of its cell membrane? You may recall that different liquid crystals possessed different phase transition temperatures. Phospholipids with the same polar head group have different transition temperatures depending upon the length of the hydrocarbon chains and the number of double bonds in the hydrocarbon chains. The shorter the hydrocarbon chains and the more double bonds present in the chains, the lower the transition temperature. Biological organisms select the proper phospholipid molecules during growth so that the transition temperature of the bilayer occurs just below the ambient temperature.

Although the physical changes due to the presence of cholesterol in cell membranes are known, the biological importance of these effects is still not understood. In general, the addition of cholesterol to a phospholipid bilayer lowers the gel to liquid crystal phase transition temperature, reduces the sharpness of this transition, and changes the rate at which molecules diffuse about the bilayer. As the cholesterol concentration rises, there also seems to be a tendency for the proteins of the cell membrane to cluster rather than to distribute themselves randomly.

It should be pointed out that cell membranes are not symmetric. That is, the outside is different from the inside. For example, some proteins lie more or less on one of the surfaces of the membrane. The proteins present on the outer surface are different from those on the inner surface. Even when the same protein is present on both surfaces, its orientation (and therefore its function) is not exactly the same on both surfaces.

There is good evidence that the phospholipid bilayer is important for the proper functioning of the proteins in the cell membrane. Proteins perform their various functions by folding up their long sequence of amino acids in a very specific way. An extremely slight rearrangement of how the protein is folded is sometimes all that is necessary to stop the protein from performing a certain function. Interactions between the proteins and the phospholipid molecules affect how the amino acid sequence of a protein is folded and therefore affect how the protein functions. We know these interactions occur, because experiments show that the phospholipid molecules in the vicinity of a protein are more highly ordered and not as free to diffuse as the phospholipid molecules in the rest of the membrane.

Other Structures

The outer cell membrane is only one of the many bilayer membranes that occur in a cell. Membranes within a cell serve to compartmentalize various functional units within the cell. For example, the nucleus of a cell is surrounded by a phospholipid membrane that is very similar to the outer cell membrane. Slight differences exist between the types of phospholipids contained in these internal membranes and obviously some of the proteins present are different. For example, the energy for a cell to function is delivered to parts of the cell by a substance called *adenosine triphosphate* (ATP), which is synthesized on the internal membrane surrounding the mitochondria. Similarly, photosynthesis takes place in plant cells on the internal membrane surrounding the chloroplasts. One last example of an internal membrane is the *endoplasmic reticulum*. This is a long, flattened bilayer structure to which the *ribosomes* (where RNA is manufactured) are attached.

From this partial list of important cell structures, it becomes clear that the phospholipid bilayer is the chief means by which different structures within a cell are segregated from the rest of the cell. In this sense, this liquid crystalline structure is one of the main building blocks of specialized structures within animal and plant cells.

The functioning of specialized cells is sometimes due to liquid crystalline structures. The visual cells of vertebrates contain stacks of flat bilayer vesicles that contain light-sensitive molecules. The high sensitivity of these cells to light is achieved by the high concentration of these molecules that results from this stacked arrangement. Impulses are carried along highly elongated nerve cells, which are surrounded by a cylindrical sleeve made of *myelin*. This myelin sleeve is made of cells (with protein-containing phospholipid bilayers) that spiral around the nerve many times. Its function is to separate the nerve cells from the surrounding environment, so they can perform their function. In this sense, the myelin sleeve functions in the same manner as insulation around a wire. As mentioned in chapter 2, the forms these sleeves adopt in water when viewed under a microscope (called *myelinic forms*) were noted as early as 1850.

Many important biological functions are possible due to the ordered arrangement of microfilaments and microtubules in cells. *Microfilaments* are long slender structures consisting of certain proteins that stack together. In this sense these microfilaments are analogous to polymers, except that the "monomer" in this case is a protein molecule. The shape and movement of cells is controlled by an oriented array of these microfilaments just under the cell membrane. Muscle cells are packed with two different types of microfilaments, which arrange themselves in a way reminiscent of nematic and smectic liquid crystals. The more slender *actin* filaments lie between the thicker *myosin* filaments as shown in figure 11.3. The myosin filaments contain a massive protein that, if activated, tends to associate

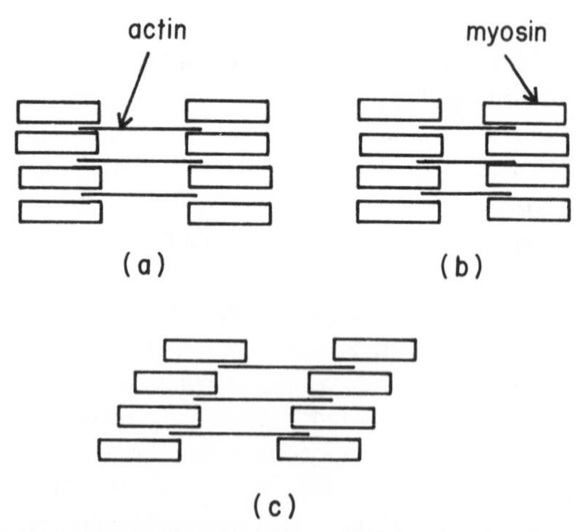

Fig. 11.3 Schematic diagram of muscle fiber, showing the actin and myosin filaments. Striated muscle is shown in its (a) relaxed and (b) contracted state. The arrangement of the filaments in smooth muscle is shown in (c).

more closely with the actin filaments. Activation of this protein on all of the myosin filaments causes contraction of the muscle cells, as depicted in figure 11.3(b). In striated muscle (for example, arms and legs), the actin filaments slide relative to the myosin filaments. Although the contraction is very rapid, it is limited in both the extent and duration of contraction. In smooth muscle (heart muscle, for example), the ordering of the actin and myosin filaments is more nematic than smectic. This arrangement is shown in figure 11.3(c). During the contraction of these cells, the actin filaments move relative to the myosin filaments and the myosin filaments move relative to each other. This produces a greater amount of contraction and the contracted state can be maintained for longer periods of time.

Microtubules are similar to microfilaments, but are larger, hollow cylinders of linearly stacked proteins instead of smaller, solid cylinders. Microtubules can be quite rigid, due to the spiral ordering of the proteins. Microtubules are thought to form the *cytoskeleton* of cells, that is, the structure responsible for cell size and shape. Microtubules are also important for some of the slow movements that occur within cells. For example, during *mitosis* (cell division), microtubules form between the chromosomes and one of the two *centrioles*. The chromosomes move along these microtubules and thus segregate with half of the chromosomes around each centriole. This process is shown in figure 11.4. The formation of a chromosome pair for each of these chromosomes and the introduction of a cell membrane then completes the process of cell division.

The arrangement of DNA in chromosomes seems to have a liquid crystalline character. The long strand of DNA winds itself back and forth in a plane perpendicular to the long axis of the chromosome, and works itself along this axis by moving from one plane to the next. The direction of the back and forth winding of the DNA rotates along the long axis of the chromosome, just as the preferred direction in a cholesteric liquid crystal rotates throughout the material. Near the side of a chromosome, the DNA strand must turn around and head back into the chromosome. As previously discussed, the double helix structure is fairly rigid, so the 180° turn cannot be extremely tight. The implication of this is that the side of a chromosome cannot be very sharply defined. Loops of the DNA strand extend into the region around the chromosome. The side of a chromosome therefore resem-

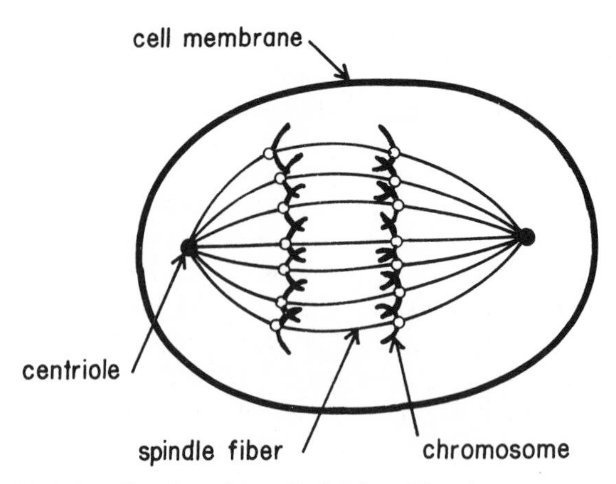

Fig. 11.4 A cell undergoing cell division. The chromosomes move along spindle fibers toward the two centrioles.

bles the interface between a liquid crystal phase and an isotropic phase. The strand of DNA is highly ordered inside the chromosome and fairly disordered outside of the chromosome. As before, however, one must keep in mind that both phases are fluid. The DNA strand is free to move or change its orientation, but when it is in the chromosome it maintains an average orientation in concert with the rest of the DNA strand. When outside the chromosome, it is much less constrained and fluctuates with little orientational or positional order on average.

There is also evidence that the liquid crystalline ordering of polymer material and proteins is the first step in the creation of very hard biological structures such as cuticle and bone. The ordering seems to be in layers, with the preferred direction in each layer rotating slightly in going from layer to layer. At first the order is liquid crystalline, but as the concentration of polymer and protein increases, the material becomes more of a gel. Crystalline bundles of parallel polymers can also form, causing the material to harden. Such a substrate seems to be necessary for minerals to begin to accumulate, thus allowing shells, bones, teeth, etc., to form.

The clotting of blood involves the production of long strands of *fibrin*, which form a close mesh to trap red blood cells. The fibrin forms from the protein *fibrogen* that is present in blood. The complex reactions that take place when a blood clot forms cause the fibrogen to polymerize and form long, ordered fibrin strands.

DISEASE

If liquid crystalline structures are essential for many biological processes, conditions that change the liquid crystalline nature of these structures can have deleterious consequences. For example, the disease multiple sclerosis is characterized by the local disintegration of the myelin sheath around nerve axons. Many factors have been implicated as possible causes of this disease, with all of them somehow having an effect on the stability of the spiral cell structure of the myelin sheath. Basic research continues to investigate the physical and chemical conditions under which such a structure is stable, in hopes of tracking down a firm explanation for this condition.

As I previously discussed, liquid crystalline structures control the size and shape of cells. Normally, red blood cells are disk shaped, with the center region thinner than the outer region. This shape possesses a large surface area, which promotes the rapid exchange of gases between the cell and the surrounding plasma. The blood of people suffering from sickle-cell anemia contains red blood cells that are distorted into a sickle shape. Not only does this shape hurt the efficiency of gas exchange, but these cells cause the viscosity of the blood to increase, thereby reducing blood flow throughout the body. The cause of sickle-cell anemia is hereditary and is due to the production of abnormal hemoglobin. This form of hemoglobin possesses a higher viscosity than normal hemoglobin, due to its tendency to form more ordered phases of the long, thin molecules. The presence of these more ordered structures tends to stretch out the red blood cells, causing them to adopt a different equilibrium shape. Either directly or indirectly, this causes a more sluggish flow of blood. An understanding of the physical characteristics of both forms of hemoglobin is certainly called for as research into sickle-cell anemia continues.

Atherosclerosis is another disease where liquid crystalline structures are implicated. In this disease, local thickening of the walls of arteries supplying blood to the heart, brain, and other vital organs obstructs the flow of blood. Investigation of the lesions that cause the thickening reveals that they contain a much higher percentage of cholesterol esters than healthy artery walls. The difference between cholesterol and cholesterol esters is important. Cholesteryl myristate is a cholesterol ester (see figure 1.10). Cholesterol is a shorter molecule, with only an HO− group instead of the $C_{13}H_{27}COO-$ group of cholesteryl myristate. Many cholesterol esters are liquid crystalline at body temperatures, whereas cholesterol has no liquid crystal phase whatsoever. Microscopic studies of diseased artery walls have revealed droplets with the classic ''brushes'' in polarized light, reminiscent of liquid crystal droplets. It is clear that some of the cholesterol esters that build up in these lesions are in the liquid crystalline phase. The reason for the buildup of cholesterol esters is not understood. There seems to be a steady increase in the amount of cholesterol present in artery walls with age. Only so much cholesterol can be present in combination with water and phospholipids before the cholesterol begins to crystallize out of solution. The creation of cholesterol ester droplets in which cholesterol can dissolve may simply be a way to contain the high amount of cholesterol without it crystallizing out. In advanced cases of atherosclerosis, the lesions actually contain crystalline cholesterol. The problem is that the rate of exchange of various substances between the lesion and the blood is fairly slow, meaning that it is difficult for the body to dissolve cholesterol or its esters once they accumulate. Knowledge of the phases present in these lesions is important to those trying to find ways to reverse the condition. As long as the cholesterol is contained in liquid or liquid crystalline phases, there is a chance that it can be removed through metabolic processes. Crystalline cholesterol presents a far greater problem, as interactions between the cholesterol molecules and surrounding substances take place at a much reduced rate. Some researchers are investigating ways to remove cholesterol from cholesterol ester droplets as a possible method to reverse the progress of the disease.

Liquid crystalline structures also seem to play a role in cancer. The microfilaments next to the cell membrane of malignant cells seem to be less ordered than in normal cells. In a normal cell, the organization of these microfilaments is coordinated around the cell surface. In this way the movements of the cell are controlled, and the cell properly responds to the presence of neighboring cells. When a normal cell becomes malignant, the coordination of the orientation of these microfilaments is lost. In some areas of the cell, the microfilaments are very disordered. These regions form irregular features that move in uncontrolled ways. The result of these uncontrolled movements of regions of the cell surface is that these cells tend to invade surrounding tissue.

The reason for this disorganization of microfilaments is unknown, but the entire cell membrane is involved. For example, one of the first signs that a cell has become malignant is that the adhesiveness between it and other cells is reduced. This shows up as small gaps between the malignant cell and surrounding cells. A reduction of the interactions between cell membranes involves the proteins on the outer surface, indicating that malignant cells have also changed the composition or structure of these outer surface proteins. The membranes of malignant cells usually contain more cholesterol than normal cells, so as we have already seen, this changes the amount of order in the bilayer structure. There is some evidence that proteins in the cell membrane of malignant cells tend to cluster more than in the membranes of normal cells. Whether this is simply a result of the change in order within abnormal cell membranes or symptomatic of something more directly involved is an open question.

Summary

What is clear is that the delicate properties of the liquid crystal phase are important to many biological processes. The normal functioning of biological structures depends on the compromise between order and disorder that characterize all liquid crystals. While this delicate response is essential for controlling all the interactions and responses taking place, it also suffers from being extremely vulnerable to anything that even slightly changes the environment. This delicate balance may be upset by the addition of a new substance, a slight change in a normal substance, a small change in the concentration of ions, or even a small change in temperature, causing a finely tuned system to malfunction. In some areas of biology, researchers are starting to ask questions about the physical state of important molecules, in hopes of uncovering important new ideas into how organisms perform necessary functions. It seems quite fitting that in many cases they will be dealing with liquid crystals, nature's delicate phase of matter.

Suggestions for Further Reading

Almost all of the literature devoted to liquid crystals has been written for specialists in the field. For this reason, all of the books listed below are at a more advanced level. Although the reader of this text should have acquired a background in liquid crystal chemistry and physics sufficient to understand the more advanced treatments used in these books, the mathematical development and scientific terminology may cause problems.

Electro-optical and Magneto-optical Principles of Liquid Crystals, by L. M. Blinov, John Wiley and Sons, 1983.
Liquid Crystals, by S. Chandrasekhar, Cambridge University Press, 1977.
The Physics of Liquid Crystals, by P. G. deGennes, Oxford University Press, 1974.
Physical Properties of Liquid Crystalline Materials, by W. H. deJeu, Gordon and Breach Science Publishers, 1980.
Textures of Liquid Crystals, by D. Demus and L. Richter, VCH Publications, 1978.
Smectic Liquid Crystals, by G. W. Gray and J.W.G. Goodby, Leonard Hill, 1984.
Handbook of Liquid Crystals, by H. Kelker and R. Hatz, Verlag Chemie, 1980.
Introduction to Liquid Crystals, by E. B. Priestley, P. J. Wojtowicz, and P. Sheng, eds., Plenum Press, 1974.
Thermotropic Liquid Crystals, Fundamentals, by G. Vertogen and W. H. deJeu, Springer-Verlag, 1988.

Index